Electromagnetic Wave Diffraction by Conducting Screens
pseudodifferential operators in diffraction problems

Electromagnetic Wave Diffraction by Conducting Screens

pseudodifferential operators in diffraction problems

A.S. Ilyinsky and Yu.G. Smirnov

Edited by Yu.V. Shestopalov

///VSP///

Utrecht, The Netherlands, 1998

VSP BV
P.O. Box 346
3700 AH Zeist
The Netherlands

First published in 1998

ISBN 90-6764-283-5

Printed in The Netherlands by Ridderprint bv, Ridderkerk.

Contents

Contents

Introduction

This monograph is devoted to an analytical investigation of vector electromagnetic problems on unclosed screens. We consider diffraction of an electromagnetic field by a perfectly conducting thin screen; by a hole in a planar, perfectly conducting screen; and in domains connected through a hole in a screen.

Interest in these problems arose long ago, and they became classical for electromagnetics. The traditional (physical) diffraction theory was created over several centuries by Huygens (1960), Fresnel (1818), Helmholtz (1859), Kirchhoff (1882), Larmore (1903), and other scientists. The Huygens principle, according to which wave propagation is caused by secondary sources, is of great importance for understanding the nature of wave processes and calculating diffraction fields. Fresnel specified the Huygens principle, taking into account the interference of spherical waves radiated by secondary sources. Kirchhoff performed a further specification of the Huygens–Fresnel principle and gave a rigorous formulation of it on the basis of the Helmholtz equation. In modern theoretical optics, methods for approximate solutions of practically all diffraction problems are based on the Huygens–Kirchhoff principle. The electromagnetic (vector) formulation of the Huygens principle was given by Kotler.

However, owing to the studies of Poincaré (1892) and Sommerfeld (1896), it became clear that problems of diffraction of electromagnetic waves are connected with certain boundary value problems of mathematical physics. Within the framework of the general setting, the problem consists of finding solutions to the Maxwell equations that satisfy certain boundary conditions, including conditions at infinity (formulated by Sommerfeld in 1912). The essence of these conditions is that all energy radiated by a source must go to infinity. In addition to this, it is necessary to take into account the behavior of fields close to the surface edges of a thin screen. Sommerfeld was the first to perform (in Refs. [18, 19]) an analytical solution to the problem of diffraction by a perfectly conducting plane. This solution allowed one to make a number of important conclusions concerning the behavior of fields in the far and near zones, in the vicinity of an edge of a thin screen, at infinity, and so on.

The first problem considered in this book, which is in fact a classical problem of electromagnetics, is the diffraction of an electromagnetic field by a bounded, perfectly conducting thin screen. Reduction to a vector integrodifferential equation on the screen [62] constitutes the most natural method of solving this problem. Such an approach is often called the method of surface currents. The idea of this method belongs to Poincaré. Note that in acoustic (scalar) problems, this method was elaborated by Rayleigh (1897). The vector integrodifferential equation on a screen was first obtained by Maue in Ref. [80] in 1949. In our notations, this equation has the form

$$Lu = grad_\tau \, A\,(div\, u) + k^2 \, A_\tau u = f, \quad x \in \Omega. \qquad (0.1)$$

Here, *div* is the operation of surface differentiation and A is the integral operator

$$Au = \int\limits_{\Omega} \frac{e^{ik|x-y|}}{|x-y|} u(y)\, ds, \tag{0.2}$$

where u is the vector field tangential to the surface Ω (the surface current density). Index τ denotes the tangential components on Ω of the corresponding vector field. The choice of spaces for solutions and right-hand sides that provides the Fredholm property (and perhaps the unique solvability) of equation (0.1) is a central problem in the studies of the solvability of this equation. In addition to this, the space of solutions must be sufficiently wide and contain all fields described by the physical model.

Investigation of equation (0.1) was started already by Maue in Ref. [80]. Later, in the fundamental monograph *Theorie der Beugung* (Springer-Verlag, 1961) by H. Hönl, A.W. Maue, and K. Westpfahl, the uniqueness theorems for equation (0.1) and for the diffraction problem were proved; the behavior of diffraction fields at infinity and in the vicinity of an edge of a screen was investigated; and analytical solutions to the problems of diffraction by a thin disk and by a sphere were obtained. It should be noted that, in the case of a thin screen, equation (0.1) was derived, with the help of the Fourier transform, in the form of an equation, which is now called the pseudodifferential equation. The authors of the monograph called it the pseudointegral equation.

In the late 1940s, Fel'd published a series of papers (see Refs. [60, 61]). He considered the problem of diffraction by a thin screen and attempted to construct a solvability theory of the boundary value diffraction problem in the space $L_1(\Omega)$ ($u \in L_1(\Omega)$). In fact, the traditional space $L_2(\Omega)$ is too narrow and does not contain solutions that have the required singularity in the vicinity of an edge of a screen (this singularity is known, for example, from the analytical solution to the problem of diffraction by a half-plane). Fel'd coordinated the choice of spaces with the behavior of fields in the vicinity of edges; however, he did not describe the space of images of the operator, defined by the left-hand side of equation (0.1).

Active applications of numerical methods (in particular, the Galerkin and the moment methods) to solving the problems of diffraction by screens of various shapes were initiated by the publication of the monograph *Field Computation by Moment Methods* (Macmillian, 1961), by R.F. Harrington. We note that, until recently, the possibility of applying any of these methods was not substantiated mathematically. Omitting the study of basic properties of operator L (separation of the main part, the Fredholm property, and so on), researchers limited themselves to the analysis of internal (computational) convergence of a method and to comparison of the results of calculations with analytical solutions. Therefore, some fundamental effects connected with characteristic properties of operator L (which will be considered below) were missed.

However, in the course of creating the methods and techniques for numerical solution of the problems of diffraction by thin screens, significant experience was accumulated. Among a great variety of publications, we note several books [3, 17, 76, 83, 98], in which the problem of diffraction by screens of different shapes is considered and, in particular, the works [5, 12, 56, 81, 82, 90] that develop numerical methods for solving the problems of diffraction by thin screens. We are not in a position to present a detailed analysis of numerical methods; we would just like to emphasize that, in the diffraction problems under consideration, the majority of calculations are performed using the moment method, the method of finite elements, and the Galerkin method with the simplest basis and probe functions (note that all these approaches are often considered as modifications of the moment method). The modern state of numerical methods is analyzed in the collection [73] of fundamental papers, published beginning from early 1920s and up to nineties. It should be noted that, in the resonant frequency range, when the wavelength is commensurable with characteristic dimensions of the screen, the appropriate methods and algorithms for efficient numerical solution to the problem of diffraction by thin screens are not available even today, when powerful computers can be applied.

The problems of diffraction by surfaces of rotation constitute a separate family. In the case of axial-symmetric excitation, they are reduced to one-dimensional equations. When excitation is not symmetric, the axial symmetry can be taken into account if the Fourier expansions of the solution and the right-hand side of equation (0.1) with respect to the azimuthal variable are performed. This procedure results in a sequence of one-dimensional equations over the generatrix of the surface of rotation. Papers and books [3, 16, 17, 65, 91, 96] deal with analytical investigations and numerical solution to the problems of diffraction by surfaces of rotation. We see, however, that the latter problem differs substantially from the general case (diffraction by a screen of arbitrary shape), because, for a surface of rotation, the diffraction problem may be reduced to one-dimensional equations.

Grinberg proposed, in Refs. [9, 10], a procedure of the transition from the vector integrodifferential equation (0.1) to a vector integral equation on the screen. He did not consider the solvability of the problem of diffraction by a planar screen. The corresponding method presupposes the solution of two additional boundary value problems for the Helmholtz equation, so that one of these problems is solved in the general form [17]. We think that such an approach does not simplify the solution.

Together with numerous attempts to solve equation (0.1) rigorously, several approximate approaches have been proposed for the problem of diffraction by a thin screen. Among them, we mention asymptotic methods [1, 2, 19, 55, 59]. We will not go into a detailed discussion of asymptotic methods for solving the problem of diffraction by unclosed surfaces. Their common disadvantage is that the accuracy of asymptotic solution and the limits of its applicability are still not clear. Note that the latter is especially important in the resonant frequency range, when the excitation wavelength is commensurable with characteristic dimensions of the surface.

Thus, in the mathematical theory of diffraction, the situation is formed, when problems are solved, by means of a great number of approximate numerical methods, for the simplest surfaces. Many particular cases are considered (for example, surfaces of rotation), but the general solvability theory has not been constructed. Here, we assume that this solvability theory includes the results of the potential theory, that is, the existence and uniqueness theorems for boundary value problems and equations over the screen (in appropriate spaces), theorems that justify the possibility of representing the solutions to boundary value problems in the form of vector potentials, theorems about discontinuities of limiting values, and so on.

There exist two classes of problems which are very close to problems of the electromagnetic wave diffraction by thin screens: vector problems of diffraction by closed, perfectly conducting surfaces and (scalar) problems of the acoustic wave diffraction by unclosed surfaces.

The first class of problems differs from that considered in this book by the fact that we study diffraction not by closed, but by unclosed surfaces. The vector nature of problems is preserved, because the boundary value problems are formulated for the Maxwell equations. The general solvability theory for electromagnetic problems of diffraction was constructed in the late sixties. Müller, who proved in Ref. [86] the existence and uniqueness theorems, managed to give a certain completeness to this theory. Owing to the results of Müller, the theory of electromagnetic wave diffraction by closed surfaces could be compared, with respect to its internal logic, with the classical potential theory. The modern state of the solvability theory for this class of problems is set forth by Colton and Kress in Ref. [27]. The solvability of the boundary value problems is proved by the reduction to a Fredholm integral equation of the second kind over the surface and is based on the properties of the corresponding vector potential [27]. The equation is considered in the Hölder spaces. Unfortunately, this technique cannot be applied to problems of diffraction by unclosed surfaces. In fact, according to the theorem that specifies the properties of limiting values, the vector potential takes different values on different sides of Ω, which contradicts the field continuity. Therefore, for unclosed surfaces, one can only derive the equation of the first kind (according to the traditional terminology). Such an equation was considered in Ref. [27]; however, its solvability was established on the basis of reduction to a known Fredholm integral equation of the second kind.

The problems of acoustic wave diffraction by unclosed surfaces constitute the second class of problems. Although these problems are scalar, they reveal some features of the problems formulated on a manifold with an edge. The solvability theory has been constructed recently in Refs. [70, 74, 75, 84, 87, 88, 94, 95]. Note that, for problems of acoustic wave diffraction by closed surfaces, similar results were already obtained in the early 1930s [29, 30], and the state-of-the-art is presented in Refs. [45, 89]. Here, the theory of pseudodifferential operators (PDOs) acting in the Sobolev spaces has become a main tool that enables one to advance in the studies of acoustic problems of diffraction by unclosed surfaces.

To date, basic branches of the general theory of PDOs are completed and set forth in the papers and books by Kohn and Nirenberg [78], Eskin [67], Shubin [66], Egorov [13, 14], Plamenevskii [37], Taylor [57], Rempel and Schulze [40], and others. Wendland was perhaps the first who started, in Ref. [95], a systematic treatment of the diffraction problem using the general theory of PDOs. He considered two-dimensional scalar problems of diffraction by thin screens and developed the corresponding solvability theory. Stephan [94] generalized these results for the case of bounded screens in R^3 with smooth edges. Note that, within the framework of the general theory of PDOs applied for scalar problems, such a generalization is comparatively easy. Screens with corner points were considered in Refs. [84, 88], where the orders of the corner singularities of solutions were calculated and the Sobolev weight classes for solutions were introduced and described. These studies continue the pioneering works of Kondrat'ev [28] and Plamenevskii [37], who examined the behavior of solutions in the vicinity of conical points.

In this book, we will solve the problems of diffraction in the case of unclosed surfaces by considering PDOs on manifolds with edges that act in the cross sections of vector bundles over Ω. The Sobolev vector spaces will be chosen in order to satisfy physical requirements of the diffraction problem.

The second problem considered in this book is the problem of diffraction of an electromagnetic field by a hole in a planar, perfectly conducting screen. This problem is dual with respect to the problem of diffraction by a bounded planar screen and is reduced to the same vector integrodifferential equation (0.1). The study of the wave diffraction by a hole is also of great theoretical and practical interest (see Refs. [42, 71, 72] and the lists of references herein). The duality principle, which establishes a connection between the problems of diffraction by a hole in a planar screen and by a bounded screen, was substantiated by Pistol'kors in 1944 and specified by Fel'd in 1960 (see Ref. [60]). The duality theorem is an analogue of the Babinet theorem, which is known in physical optics (this theorem couples, within the framework of the scalar Hyugens principle, the diffraction phenomena for reciprocal screens).

The third problem of diffraction by a partially shielded layer differs from the previous one by the presence of an additional shielding, perfectly conducting plane placed in one of the half-spaces. This problem is usually solved with the help of the Green's function of the Helmholtz equation for the layer. The Green's functions are studied in detail in Refs. [79, 85]. The main difficulty here is the formulation of conditions at infinity. The appropriate conditions were formulated first by Sveshnikov in Ref. [46]. Werner [85] generalized these conditions, and now, they are often called the Sveshnikov–Werner partial radiation conditions at infinity. The problem of diffraction by a partially shielded layer is also reduced to an equation over a hole.

Electromagnetic wave diffraction by an aperture in a screen that separates a half-space and a semi-infinite rectangular waveguide constitutes the fourth problem. This diffraction problem was also the subject of numerous studies [41, 68, 77]. Here, the field in the cylindrical domain cannot be expressed in the form of potentials if only one (scalar) Green's function is applied. In Ref. [41], it was proposed to represent a solution using two Green's functions with mixed boundary conditions and the Sveshnikov–Werner radiation conditions at infinity. Analysis of the Green's functions is connected with investigations of the convergence of series in eigenfunctions of the Laplacian. The study of convergence of such expansions is completed by Il'in in Refs. [20–22].

The last three problems describe the diffraction in domains connected through a hole in a screen and are reduced to the same integrodifferential equation over the hole. We have chosen these problems, because, for them, one can construct explicit representations of the Green's functions of the corresponding domains and reveal three types of the behavior of solutions at infinity, namely, in the half-space, in a layer, and in a cylindrical domain. In addition to this, various practical problems (see [73, 77]) call for creating efficient methods of solution. Note that the solvability theory for these problems is not constructed.

In this book, we study the solvability of three-dimensional vector electromagnetic problems on unclosed surfaces. We prove the following statements.

(i) The existence and uniqueness theorems (in appropriate spaces) of the boundary value problem for the Maxwell equations.

(ii) The possibility of representing the solutions to boundary value problems in the form of vector potentials or in the form of similar expressions obtained with the help of the corresponding Green's functions.

(iii) Reduction of the boundary value problem to an equation on a manifold with an edge (when the dimensionality of the problem is reduced).

(iv) Theorems about discontinuities of vector potentials or other representations of solutions to boundary value problems.

(v) Theorems about the solvability of the equation considered on a manifold in appropriate spaces.

(vi) Regularity theorems that specify the smoothness of solutions, including the behavior in the vicinity of edges and corner points of the manifold.

(vii) The dependence of solutions to the equation and the boundary value problem on parameters.

We study the problems of diffraction by unclosed surfaces on the basis of the theory of PDOs that act in the Sobolev spaces of the cross sections of vector bundles [15, 35, 40, 57]. The diffraction problems are analyzed according to the single scheme. Each problem is reduced to a pseudodifferential equation on a manifold with an edge Ω (a screen or a hole in a screen). The corresponding PDO $L : H_1 \to H_2$ is considered in a specially chosen Hilbert space. We use the standard method developed in Refs. [25, 26] to prove that L is a bounded Fredholm operator with the zero index. L is represented as a sum of a (bounded) invertible operator S and a (bounded) compact operator K, $L = S + K$. The solution to the boundary value problem is sought in the form of vector potentials (or similar representations obtained with the help of the Green's functions). The uniqueness of the solution to the boundary value problem yields the uniqueness for the corresponding pseudodifferential equation $Lu = f$, $u \in H_1$. The solvability of the equation for the arbitrary right-hand side $f \in H_2$, which yields the solvability of the initial boundary value problem, follows then from the Fredholm alternative.

The results set forth in this book are published in Refs. [24, 48–54].

Chapter 1

Diffraction by cylindrical screens

To date, the solvability theory for the two-dimensional problems of the electromagnetic and acoustic wave diffraction by thin cylindrical screens is completed. In this chapter, we summarize the main results concerning the solvability of two-dimensional diffraction problems. The majority of them are well-known, and we will limit ourselves to a brief survey, assuming that details can be found in original monographs. This chapter is an introduction to the essentials related to three-dimensional diffraction problems.

The solvability theory for the problems of diffraction by cylindrical screens is included as a standard of the completed diffraction theory and to demonstrate the differences between two-dimensional and three-dimensional problems. We note also that the solution techniques for these problems are often similar.

In Section 1.1 we focus on a mathematical statement of the problem of diffraction by cylindrical screens: solutions are represented as single-layer and double-layer potentials and pseudodifferential equations on the screen are derived. The solvability theory is constructed for pseudodifferential equations and boundary value problems in the Sobolev spaces.

Section 1.2 is devoted to the problems of diffraction by closed screens. Pseudodifferential equations are solved numerically with the help of the Galerkin method.

In Section 1.3 we consider the diffraction problems for unclosed screens beginning with an important particular case of planar screens, when the spaces of solutions and images and pseudodifferential operators (PDOs) can be described in terms of Fourier transforms. The Galerkin methods are applied in the weighted Sobolev spaces with an account for singularities of solutions in the vicinity of a screen edge.

1.1 Statement of the problems of diffraction by cylindrical screens

Consider a system composed of a finite number of cylindrical screens S_j, where S_j coincides with the cylindrical surface formed by a simple, planar, closed or unclosed smooth curve Γ_j of a finite length that belongs to class C^∞ (note that according to Ref. [24], the results presented below are valid for piece-wise smooth cylindrical screens). We introduce the following notation:

$$\Gamma = \bigcup_j \Gamma_j, \quad \overline{\Gamma}_i \bigcap \overline{\Gamma}_j = \emptyset \quad \text{for} \quad i \neq j,$$

where

$$\partial\Gamma = \bigcup_j (\overline{\Gamma}_j \setminus \Gamma_j)$$

are the endpoints of Γ (these points do not belong to Γ and $\partial\Gamma \bigcap \Gamma = \emptyset$). Let G_0 be a union of internal domains bounded by closed curves Γ_j, and we set $G_0 = \emptyset$ if these domains are absent.

We assume that the Ox_3 axis is directed along the generatrix of surfaces S_j and the incident field does not depend on x_3. Then both the acoustic and the electromagnetic problems of diffraction by screens S_j are reduced [62] to determination of the scalar function u (the scattered field) that satisfies the homogeneous Helmholtz equation

$$\Delta u + k^2 u = 0 \quad \text{in} \quad \mathrm{R}^2 \setminus (\overline{\Gamma} \bigcup G_0), \tag{1.1}$$

the Dirichlet,

$$u|_{\Gamma} = -f \tag{1.2}$$

or the Neumann,

$$\left.\frac{\partial u}{\partial n}\right|_{\Gamma} = -g \tag{1.3}$$

boundary value conditions on Γ, and the Sommerfeld radiation conditions at infinity

$$u = O(r^{-1/2}), \quad \frac{\partial u}{\partial r} - iku = o(r^{-1/2}) \tag{1.4}$$

for $r := (x_1^2 + x_2^2)^{1/2} \to \infty$. The Dirichlet problem corresponds to a soft screen in acoustics (or to the case of E-polarization in electrodynamics), and the Neumann problem, to a hard screen (or to the case of H-polarization). Here, k is the free-space wavenumber and $\Im k \geq 0$, $k \neq 0$.

In addition to the above conditions, the field u must satisfy the requirement that provides the finiteness of energy in every bounded spatial domain:

$$u \in H^1_{loc}(\mathrm{R}^2 \setminus \overline{G_0}), \tag{1.5}$$

that is,

$$\int_{G\setminus G_0} \left(|\nabla u|^2 + |u|^2 \right) dx < \infty$$

for every bounded domain G.

Let

$$\Gamma_\delta = \{x : dist(x, \partial\Gamma) < \delta\}$$

be a δ-vicinity of the endpoints of curve Γ, and Λ, a closed, smooth C^∞-curve that contains Γ ($\Gamma \subset \Lambda$) and divides the plane into bounded, G_1, and unbounded, G_2 domains. We will assume that n is the normal vector external with respect to G_1.

Let us formulate the definitions of the Sobolev spaces that will be used in this book.

Consider Λ as a closed manifold and denote by $U = \{U_\nu\}$ a finite cover of Λ by the coordinate vicinities and by $\{\Phi_\nu\}$, the partition of unity subordinated to the cover of U. For every $v \in C^\infty(\Gamma)$, we introduce the functions $v_\nu = \Phi_\nu v$ defined in local coordinated in a certain interval of the real line and set

$$\|v\|_s^2 = \sum_\nu \|v_\nu\|_s^2, \quad s \in \mathrm{R}$$

for every $v \in C^\infty(\Lambda)$. The norm in the space $H^s(\mathrm{R}^1)$ is defined as

$$\|v_\nu\|_s^2 = \int\limits_{-\infty}^{\infty} \left(1+|\xi|^2\right)^s |\widehat{v}_\nu(\xi)|^2\, d\xi,$$

where $\widehat{v}_\nu$ is the Fourier transform of function v_ν. By $H^s(\Lambda)$ we will denote the completion of $C^\infty(\Lambda)$ with respect to the $\|\cdot\|_s$-norm. Note that if the coordinate vicinities and the partition of unity are chosen in a different manner, one obtains an equivalent norm; therefore, the definition of $H^s(\Lambda)$ is correct.

Let $\Gamma \subset \Lambda$ be a submanifold of Λ (a part of curve Λ); we set

$$H^s(\Gamma) := \{v|_\Gamma : v \in H^s(\Lambda)\}$$

and

$$\tilde{H}^s(\Gamma) := \left\{v \in H^s(\Lambda) : supp\, v \subset \overline{\Gamma}\right\}.$$

Spaces $H^s(\Gamma)$ and $\tilde{H}^{-s}(\Gamma)$ are antidual, for all $s \in \mathrm{R}$, with respect to sesquilinear form $\int\limits_\Gamma v\overline{w}\, dl$. The space $\tilde{H}^s(\Gamma)$ may be constructed as a closure of $C_0^\infty(\Gamma)$ with respect to the norm $\|\cdot\|_s$. For a detailed description of the properties of spaces $H^s(\Lambda)$, $H^s(\Gamma)$, and $\tilde{H}^s(\Gamma)$, one can refer, for example, to Refs. [40, 57].

Let Γ be an open smooth curve of a finite length with the endpoints p_i. If

$$\varphi = \varphi_0 + \sum_i \alpha_i \rho_i^{-1/2} \chi_i, \tag{1.6}$$

where $\alpha_i \in \mathrm{R}$, $\varphi_0 \in \tilde{H}^\tau(\Gamma)$, $\tau < 1$, then we will write

$$\|\varphi\|_{Z^\tau(\Gamma)} := \begin{cases} \sum\limits_i |\alpha_i| + \|\varphi_0\|_{\tilde{H}^\tau(\Gamma)}, & 0 \le \tau < 1, \\ \|\varphi\|_{\tilde{H}^\tau(\Gamma)}, & \tau < 0, \end{cases}$$

denoting by $Z^\tau(\Gamma)$ the corresponding space, so that $\varphi \in Z^\tau(\Gamma)$. Here, ρ_i is the distance to the endpoint p_i and χ_i, $0 \le \chi_i \le 1$, is the cut function that belongs to C^∞ and equals unity in the vicinity of p_i and zero, in the vicinities of all other endpoints.

For functions φ admitting representation in the form (1.6), for which

$$\varphi_0 = \sum_i \beta_i \rho_i^{1/2} \chi_i + \varphi_1, \tag{1.7}$$

where $\beta_i \in \mathrm{R}$ and $\varphi_1 \in \tilde{H}^\tau(\Gamma)$ for $\tau < 2$, we define

$$\|\varphi\|_{Z^\tau(\Gamma)} := \begin{cases} \sum\limits_i (|\alpha_i| + |\beta_i|) + \|\varphi_1\|_{\tilde{H}^\tau(\Gamma)}, & 1 \le \tau < 2, \\ \|\varphi\|_{\tilde{H}^\tau(\Gamma)}, & \tau < 1, \end{cases}$$

and denote by $Z^\tau(\Gamma)$ the corresponding space, so that $\varphi \in Z^\tau(\Gamma)$, $\tau < 2$. A number of statements concerning the properties of the space $Z^\tau(\Gamma)$ can be found in Ref. [95]. For $\tau \ge 2$, the spaces $Z^\tau(\Gamma)$ can be introduced in a similar manner by separating the singularities $\rho^{3/2}$, $\rho^{5/2}$, and so on.

Let us introduce the following operators: γ_0, of the trace on Λ; γ_1, of the trace of the normal derivative on Λ (on each side of the curve); q, of the zero continuation of a function from Γ to Λ; and p, of the zero restriction of a function from Λ to Γ:

$$\gamma_0 : u \to u|_\Lambda : H^1_{loc}(\mathrm{R}^2) \to H^{1/2}(\Lambda),$$
$$\gamma_1 : u \to \left.\tfrac{\partial u}{\partial n}\right|_\Lambda : H^1_P(G) \to H^{-1/2}(\Lambda),$$
$$q : \varphi \to \tilde{\varphi} : \tilde{H}^s(\Gamma) \to H^s(\Lambda),$$
$$p : \varphi \to \varphi|_L : H^s(\Lambda) \to H^s(\Gamma),$$

where $G = G_1$ or $G = G_2$ and $H^1_P(G) = \{u \in H^1(G) : \Delta u \in L_2(G)\}$. According to Ref. [75], all operators are continuous on the corresponding pair of spaces; therefore, it is sufficient to define them for smooth functions (in an obvious manner) and to extend then to the whole space using continuation.

In (1.2) and (1.3), the equality means the equality of elements from the spaces $H^{1/2}(\Gamma)$ and $H^{-1/2}(\Gamma)$, and (1.3) must hold on both sides of Γ. It is well known [34] that solutions to the homogeneous Helmholtz equation from $H^1_{loc}(\mathrm{R}^2 \setminus (\overline{\Gamma} \bigcup G_0))$ are infinitely differentiable in $\mathrm{R}^2 \setminus (\overline{\Gamma} \bigcup G_0)$; therefore, from the very beginning, one can assume that $u \in C^2(\mathrm{R}^2 \setminus (\overline{\Gamma} \bigcup G_0))$ and understand (1.1) in a usual sense. If the field sources are situated outside the screens, the functions f and g, which simulate the trace of the incident field and its normal derivative on Γ, will be infinitely differentiable on Γ and $f, g \in C^\infty(\overline{\Gamma})$. Then, if $\Gamma' \subset \Gamma$ is a smooth part of a C^∞-curve, one can use the regularity theorem [34] to show that solution u is infinitely differentiable up to Γ' (on each side of the curve, if $L' \bigcap \partial G_0 = \emptyset$), and

$$u \in C^2(\mathrm{R}^2 \setminus (\overline{\Gamma} \bigcup G_0)) \bigcap_{\delta>0} C^1(\overline{G_1 \setminus (\Gamma_\delta \bigcup G_0)}) \bigcap_{\delta>0} C^1(\overline{G_2 \setminus \Gamma_\delta}),$$

if all Γ_j belong to C^∞. In Ref. [95], diffraction problem (1.1)–(1.5) is investigated, in the generalized setting, for a single smooth screen.

Consider the uniqueness of the problems formulated above. We will prove that for $\Im k \geq 0$, $k \neq 0$, the Dirichlet and Neumann problems with homogeneous boundary conditions have only trivial solutions. Denote by $[\,\cdot\,]$ the difference of the limiting values of a function taken from different sides of a curve. (1.1)–(1.3) yields the conjugation problem for u:

$$\Delta u + k^2 u = 0 \quad \text{in} \quad G_2 \bigcup (G_1 \setminus G_0),$$

$$[u]_{\Lambda\setminus\overline{\Gamma}} = \left[\frac{\partial u}{\partial n}\right]_{\Lambda\setminus\overline{\Gamma}} = 0,$$

$$u|_\Gamma = 0 \quad \text{or} \quad \left.\frac{\partial u}{\partial n}\right|_\Gamma = 0,$$

with radiation condition (1.4) for $r \to \infty$. Denote by $B_R = \{\mathrm{x} : |\mathrm{x}| < R\}$ a circle of radius R such that $\Lambda \subset B_R$. Applying the second Green's formula in domains $G_1 \setminus G_0$, $G_2 \bigcap B_R$ to functions u and $\overline{u}$ and adding up the results, we obtain

$$\int_{\partial B_R} \left(\frac{\partial u}{\partial r}\overline{u} - \frac{\partial \overline{u}}{\partial r}u\right) dl = -4ik'k'' \int_{B_R} |u|^2 \, d\mathrm{x}, \quad k = k' + ik''.$$

Integrals over Λ vanish by virtue of the boundary conditions and the conjugation conditions. Here, applicability of the Green's formula follows from the fact that u belongs to $H^1_{loc}(\mathrm{R}^2 \setminus G_0)$ and contour Λ is smooth [75]. The standard analysis of the obtained identity (see Ref. [23]) shows that $u \equiv 0$ in G_2, and the conjugation conditions and the analytical property of u in $\mathrm{R}^2 \setminus (\overline{\Gamma} \bigcup G_0)$ yields $u \equiv 0$ in $G_1 \setminus G_0$. Thus, uniqueness of the solution to problems (1.1)–(1.5) is proved .

We will look for the unique solution to problems (1.1)–(1.5) in the form of potentials

$$u(\mathrm{x}) = K_0(q\varphi) = -\frac{i}{4}\int\limits_{\Gamma} H_0^{(1)}(k|\mathrm{x} - y|)\varphi(y)\,dl, \tag{1.8}$$

$$\varphi \in \tilde{H}^{-1/2}(\Gamma)$$

for the Dirichlet problem and

$$u(\mathrm{x}) = K_1(q\psi) = \frac{i}{4}\int\limits_{\Gamma} \frac{\partial}{\partial n_y} H_0^{(1)}(k|\mathrm{x} - y|)\psi(y)\,dl, \tag{1.9}$$

$$\psi \in \tilde{H}^{1/2}(\Gamma)$$

for the Neumann problem. Here, $H_0^{(1)}(z)$ is the Hankel function of the first kind. For the case of a closed contour, potentials (1.8) and (1.9) are investigated in Ref. [75]. Continuing φ and ψ as zero functions from Γ to Λ and performing the restriction on Γ, we obtain, by virtue of the continuity of operators q and p and the results of Ref. [75], that

$$\varphi = \left[\frac{\partial u}{\partial n}\right]_{\Gamma} \text{ and } [u]_{\Gamma} = 0 \text{ for the Dirichlet problem}$$

and

$$\psi = [u]_{\Gamma} \text{ and } \left[\frac{\partial u}{\partial n}\right]_{\Gamma} = 0 \text{ for the Neumann problem.}$$

We prove also that

$$u \in H^1_{loc}(\mathrm{R}^2 \setminus \overline{G_0})$$

for all

$$\varphi \in \tilde{H}^{-1/2}(\Gamma), \quad \psi \in \tilde{H}^{1/2}(\Gamma)$$

in (1.8) and (1.9). Here,

$$[u]_{\Gamma} = p\left(\gamma_0\left(u|_{G_2}\right) - \gamma_0\left(u|_{G_1}\right)\right),$$

$$\left[\frac{\partial u}{\partial n}\right]_{\Gamma} = p\left(\gamma_1\left(u|_{G_2}\right) - \gamma_1\left(u|_{G_1}\right)\right).$$

Since solution u is chosen in the form of potentials (1.8) and (1.9), equation (1.1) and conditions (1.4) and (1.5) hold for all φ and ψ and we have to satisfy only boundary conditions (1.2) or (1.3), which yields the equations on screens Γ:

$$D\varphi := \frac{i}{4}\int\limits_{\Gamma} H_0^{(1)}(k|\mathrm{x} - y|)\varphi(y)\,dl = f(\mathrm{x}), \quad \mathrm{x} \in \Gamma, \tag{1.10}$$

$$D = p\gamma_0 K_0 q,$$

for the Dirichlet problem and

$$N\psi := \frac{i}{4}\frac{\partial}{\partial n_x}\int_\Gamma \frac{\partial}{\partial n_y} H_0^{(1)}(k|x-y|)\psi(y)\,dl = -g(x), \quad x \in \Gamma; \tag{1.11}$$

$$N = p\gamma_1 K_1 q,$$

for the Neumann problem. The first equation has a weak (logarithmic) singularity and the second equation is hypersingular. In spite of substantial differences, both equations (1.10) and (1.11) may be considered from a single viewpoint as pseudodifferential equations.

Thus, if φ and ψ are solutions to (1.10) and (1.11), formulas (1.8) and (1.9) give the solution to diffraction problem (1.1)–(1.5).

1.2 Closed cylindrical screens

Consider the case when all curves Γ_j are closed and smooth. Then, according to Ref. [75],

$$D : H^{-1/2}(\Gamma) \to H^{1/2}(\Gamma)$$

and

$$N : H^{1/2}(\Gamma) \to H^{-1/2}(\Gamma)$$

are the Fredholm operators with zero indices. D and N are elliptic PDOs of orders -1 and $+1$, respectively.

By virtue of the regularity theorem, if the equations

$$D\varphi = f \tag{1.12}$$

and

$$N\psi = -g \tag{1.13}$$

have C^∞ right-hand sides and Γ is a C^∞ curve, the solutions φ and ψ on Γ will belong to C^∞. Equations (1.12) and (1.13) or (1.10) and (1.11) are uniquely solvable for all k: $\Im k \geq 0$, $k \neq 0$ except for a discrete set of points—characteristics numbers (CNs)—situated on the real axis with the only possible accumulation point at infinity. For equations (1.12) and (1.13), all CNs are isolated and have finite algebraic multiplicity. We will denote by $\sigma(D)$ and $\sigma(N)$ the sets of CNs for equations (1.12) and (1.13), respectively. For $k \in \sigma(D)$ or $k \in \sigma(N)$, when the Dirichlet or the Neumann problem are solved, (1.12) or (1.13) are replaced by modified equations, which are uniquely solvable for $k \in \sigma(D)$ or $k \in \sigma(N)$. Several ways of obtaining such equations are described in Refs. [27, 89]. Thus, for all values k such that $\Im k \geq 0$ and $k \neq 0$, the Dirichlet and the Neumann problems (1.1)–(1.5) are uniquely solvable. However, the solutions can be represented in the form of potentials (1.8) and (1.9) only if $k \notin \sigma(D)$ for the Dirichlet problem and $k \notin \sigma(N)$ for the Neumann problem. In the vicinities of the spectrum points, numerical solution of equations (1.12) and (1.13) is often unstable and the rate of convergence decreases (which depends on the choice of the numerical method).

For complicated contours, CNs are not known; therefore, the following specific methods are applied for obtaining the correct solution. First, (1.12) and (1.13) can be replaced by modified equations, but in this case, the kernels of integral equations become complicated and calculations are more bulky. Second, one can artificially introduce absorption, replacing a real k by $k + i\varepsilon$ with a small $\varepsilon > 0$. Third, a regularizing procedure can be applied to solving the system of linear algebraic equations.

Consider the convergence of the numerical method for solving equations (1.12) and (1.13) assuming that k is fixed and $k \notin \sigma(D)$ for (1.12) and $k \notin \sigma(N)$ for (1.13).

We begin with equation (1.12). Let $\left\{W^h\right\}_{h>0}$ be a family of subspaces of the space $H^{-1/2}(\Gamma)$ such that for every $v \in H^{-1/2}(\Gamma)$, the quantity $\|v - P^h v\|_{-1/2} \to 0$ for $h \to 0$, where P^h is an orthogonal projector on W^h in $H^{-1/2}(\Gamma)$. Consider the Galerkin method for solving equation (1.12): find an element $\varphi_h \in W^h$, for which

$$\langle D\varphi_h, v\rangle = \langle f, v\rangle, \quad \forall v \in W^h. \tag{1.14}$$

Here, brackets $\langle\cdot,\cdot\rangle$ denote the antiduality property of spaces $H^{1/2}$ and $H^{-1/2}$, that is, a continuous extension on $H^{1/2}$ and $H^{-1/2}$ of the form

$$\langle f, v\rangle = \int_\Gamma f\bar{v}\, dl. \tag{1.15}$$

If, for example, $v \in L_2(\Gamma)$, then brackets $\langle\cdot,\cdot\rangle$ may be replaced by integral (1.15).

If W^h is a finite-dimensional space, the Galerkin scheme yields a system of linear algebraic equations. Let $\{\varphi_h^i\}$ be a basis in W^h. Then, (1.14) is equivalent to

$$D^h a_h = b_h, \tag{1.16}$$

where

$$D^h = \left\{\langle D\varphi_h^i, \varphi_h^j\rangle\right\}_{i,j}, \quad b_h = \left\{\langle f, \varphi_h^j\rangle\right\}_j^T,$$
$$a_h = \{a_h^i\}_i^T, \quad \varphi_h = \sum_i a_h^i \varphi_h^i.$$

The following statement is valid [75].

Statement 1 *Let $k \notin \sigma(D)$. Then, there exists $h_0 > 0$ such that for all $0 < h < h_0$, equation (1.14) has the unique solution $\varphi_h \in W^h$. For $h \to 0$, φ_h converge to the unique solution φ in the quasi-optimal sense; that is, there exists a constant C such that for all $0 < h < h_0$, the following estimate is valid:*

$$\|\varphi - \varphi_h\|_{-1/2} \le C \inf_{v \in W^h} \|\varphi - v\|_{-1/2} = C E_{-1/2}(\varphi, W^h). \tag{1.17}$$

Using general expression (1.17), one can estimate the rate of convergence when W^h is known. Assume, for example, that W^h is the space of piece-wise constant elements (zero-order splines) given on certain regular grid depending on the parameter h equal to the maximal step of the grid. Then, (1.17) yields the estimate [75]

$$\|\varphi - \varphi_h\|_{-1/2} \le C h^{1/2} \|f\|_1 \tag{1.18}$$

for $f \in H^1(\Gamma)$.

In order to estimate the value $E_{-1/2}(\varphi, W^h)$ of the best approximation of $\varphi \in H^{-1/2}(\Gamma)$ by elements from $W^h \subset H^{-1/2}(\Gamma)$, one can apply a developed theory (see Ref. [39] and the list of references presented in this book), introducing the parametrization of curve Γ and proceeding to the analysis on an interval.

Consider equation (1.13): $N\psi = -g$. Let $\left\{W^h\right\}_{h>0}$ be a family of subspaces of the space $H^{1/2}(\Gamma)$ such that for every $v \in H^{1/2}(\Gamma)$, $\left\|v - Q^h v\right\|_{1/2} \to 0$ as $h \to 0$, where Q^h is an orthogonal projector on W^h in $H^{1/2}(\Gamma)$. The Galerkin method for solving equation (1.13) consists in the following: find an element $\psi_h \in W^h$ such that

$$\langle N\psi_h, v\rangle = -\langle g, v\rangle, \quad \forall v \in W^h. \tag{1.19}$$

Here, brackets $\langle\cdot,\cdot\rangle$ denote the antiduality property of spaces $H^{-1/2}$ and $H^{1/2}$, that is, a continuous extension on $H^{-1/2}$ and $H^{1/2}$ of the form

$$\langle g, v\rangle = \int_\Gamma g\bar{v}\, dl \tag{1.20}$$

If $g \in L_2(\Gamma)$, the brackets $\langle\cdot,\cdot\rangle$ may be replaced by integral (1.20).

If W^h is a finite-dimensional space, the Galerkin scheme (1.19) yields a system of linear algebraic equations. Let $\{\psi_h^i\}$ be a basis in W^h. Then, (1.19) is equivalent to

$$N^h a_h = -b_h, \tag{1.21}$$

where

$$N^h = \left\{\langle N\psi_h^i, \psi_h^j\rangle\right\}_{i,j}, \quad b_h = \left\{\langle g, \psi_h^j\rangle\right\}_j^T,$$
$$a_h = \{a_h^i\}_i^T, \quad \psi_h = \sum_i a_h^i \psi_h^i.$$

The following statement is valid [75].

Statement 2 *Let $k \notin \sigma(N)$. Then there exists $h_0 > 0$ such that for all $0 < h < h_0$ equation (1.19) has a unique solution $\psi_h \in W^h$. For $h \to 0$, ψ_h converge to the unique solution ψ of equation (1.13) in the quasi-optimal sense; that is, there exists a constant C such that for all $0 < h < h_0$, the following estimate is valid:*

$$\|\psi - \psi_h\|_{1/2} \le C \inf_{v\in W^h} \|\psi - v\|_{1/2} = CE_{1/2}(\psi, W^h). \tag{1.22}$$

Similar to the case of the Dirichlet problem, one can estimate the rate of convergence when W^h is known using general expression (1.22). Note, however, that piece-wise constant elements do not belong to the space $H^{1/2}(\Gamma)$; therefore, it is necessary to choose smoother functions ψ_h^i.

1.3 Unclosed cylindrical screens

Assume that all Γ_j are unclosed smooth curves. Before to proceed to the analysis of the general case of the diffraction problem on Γ, consider a particular case of planar screens situated in one plane, when

$$\Gamma = \{\mathrm{x} : \mathrm{x}_2 = 0,\ a_{2j-1} < \mathrm{x}_1 < a_{2j},\ j = 1, \ldots, J;\ a_i < a_j\ (i < j)\}.$$

In this case, D and N are convolution-type operators on the line R^1. Calculating the Fourier transforms of the kernels of these operators and using the convolution theorem and the properties of the Fourier transform, we rewrite equations (1.10) and (1.11) as

$$D\varphi \equiv \int_{-\infty}^{+\infty} \frac{1}{\sqrt{\xi^2 - k^2}} e^{i\mathrm{x}_1\xi} \widehat{\varphi}(\xi)\, d\xi = f(\mathrm{x}_1), \quad \mathrm{x}_1 \in \Gamma, \tag{1.23}$$

and

$$N\psi \equiv \int\limits_{-\infty}^{+\infty} \sqrt{\xi^2 - k^2} e^{ix_1\xi} \widehat{\psi}(\xi)\, d\xi = g(x_1), \quad x_1 \in \Gamma. \tag{1.24}$$

Here, $\widehat{\varphi}$ and $\widehat{\psi}$ denote the Fourier transforms of φ and ψ. We will look for solutions φ and ψ in the spaces of generalized functions (distributions) $\varphi \in \tilde{H}^{-1/2}(\Gamma)$ and $\psi \in \tilde{H}^{1/2}(\Gamma)$. In the case under consideration, these spaces can be described in terms of the Fourier transform:

$$\tilde{H}^s(\Gamma) = \left\{ u : (1 + |\xi|)^s \widehat{u}(\xi) \in L_2(\mathrm{R}^1),\ supp\, u \subset \overline{\Gamma} \right\}.$$

The left-hand sides of equations (1.23) and (1.24) define the boundary integrodifferential operators

$$D : \tilde{H}^{-1/2}(\Gamma) \to H^{1/2}(\Gamma)$$

and

$$N : \tilde{H}^{1/2}(\Gamma) \to H^{-1/2}(\Gamma),$$

where $H^s(\Gamma)$ is a restriction of $H^s(\mathrm{R}^1)$ on Γ. Since

$$\begin{array}{lcl} (\xi^2 - k^2)^{-1/2} & = & |\xi|^{-1} + O(|\xi|^{-3}), \\ (\xi^2 - k^2)^{1/2} & = & |\xi| \quad + O(|\xi|^{-1}) \end{array}$$

for $|\xi| \to \infty$, then, according to Ref. [67], D and N are the Fredholm PDOs with the zero index and the orders -1 and $+1$, respectively. These operators are invertible by virtue of the estimates

$$\left|(D\varphi, \varphi)_{L_2}\right| \geq \frac{1}{\sqrt{2}} \int\limits_{-\infty}^{+\infty} \left|\xi^2 - k^2\right|^{-1/2} |\widehat{\varphi}(\xi)|^2\, d\xi > 0, \quad \varphi \neq 0$$

and

$$\left|(N\psi, \psi)_{L_2}\right| \geq \frac{1}{\sqrt{2}} \int\limits_{-\infty}^{+\infty} \left|\xi^2 - k^2\right|^{1/2} |\widehat{\psi}(\xi)|^2\, d\xi > 0, \quad \psi \neq 0.$$

These estimates prove that the kernels of these operators contain only the zero element: $ker\, D = \{0\}$ and $ker\, N = \{0\}$. Thus, owing to the Fredholm alternative, equations (1.23) and (1.24) or (1.10) and (1.11) are uniquely solvable.

Consider the general case of a system of screens of arbitrary shape Γ. Operators

$$D : \tilde{H}^{-1/2}(\Gamma) \to H^{1/2}(\Gamma)$$

and

$$N : \tilde{H}^{1/2}(\Gamma) \to H^{-1/2}(\Gamma),$$

are bounded, and their Fredholm property follows from the Gårding inequality [95]

$$\begin{array}{lcl} \Re\langle (D + K_D)\varphi, \varphi\rangle & \geq & \lambda_D \left\|\varphi\right\|^2_{-1/2}, \quad \forall \varphi \in \tilde{H}^{-1/2}(\Gamma), \\ \Re\langle (N + K_N)\psi, \psi\rangle & \geq & \lambda_N \left\|\varphi\right\|^2_{1/2}, \quad \forall \psi \in \tilde{H}^{1/2}(\Gamma) \end{array}$$

with certain compact operators

$$\begin{array}{rl} K_D : & \tilde{H}^{-1/2}(\Gamma) \to H^{1/2}(\Gamma), \\ K_N : & \tilde{H}^{1/2}(\Gamma) \to H^{-1/2}(\Gamma). \end{array}$$

Equations (1.10) and (1.11) are uniquely solvable for arbitrary right-hand sides $f \in H^{1/2}(\Gamma)$ or $g \in H^{-1/2}(\Gamma)$, since the corresponding homogeneous equations have only trivial solutions (if we assume that φ and ψ are nontrivial solutions of (1.10) and (1.11) for $f = 0$ and $g = 0$, then formulas (1.8) and (1.9) yield nontrivial solutions of homogeneous problems (1.1)–(1.5), which contradicts the uniqueness theorem).

Thus, in the case of unclosed screens, equations (1.10) and (1.11) has a unique solution for all k: $\Im k \geq 0$, $k \neq 0$. Therefore, here, it is not necessary to construct any modified equations or to apply regularizing procedures connected with the presence of a nonempty spectrum.

When the diffraction problems are considered, it is important to investigate the asymptotic behaviour of solutions in the vicinities of the endpoints of $\partial\Gamma$ (edges of screens). Let f and g be smooth functions, e.g., $f, g \in C^\infty(\overline{\Gamma})$. Applying the known results concerning regularity of solutions of elliptic equations, it is easy to show that φ and ψ are smooth functions on Γ (they belong to C^∞ if all $\Gamma_j \in C^\infty$) and have singularities

$$\varphi = O(\rho^{-1/2}), \quad \rho \to 0, \tag{1.25}$$

$$\psi = O(\rho^{1/2}), \quad \rho \to 0, \tag{1.26}$$

in the vicinities of the endpoints of $\partial\Gamma$ (ρ is the distance to an endpoint of the screen); note that estimates (1.25) and (1.26) are accurate with respect to the order.

Consider the Galerkin method for solving equations (1.10) and (1.11). Let

$$D\varphi = f$$

and $\left\{W^h\right\}_{h>0}$ be a family of subspaces of the space $\tilde{H}^{-1/2}(\Gamma)$ such that for every $v \in \tilde{H}^{-1/2}(\Gamma)$, $\|v - P^h v\|_{-1/2} \to 0$ for $h \to 0$, where P^h is an orthogonal projector on W^h in $\tilde{H}^{-1/2}(\Gamma)$. It is necessary to find $\varphi_h \in W^h$ such that

$$\langle D\varphi_h, v\rangle = \langle f, v\rangle, \qquad \forall v \in W^h. \tag{1.27}$$

Here, brackets $\langle\cdot,\cdot\rangle$ denote the antiduality property of spaces $H^{1/2}$ and $\tilde{H}^{-1/2}(\Gamma)$, that is, a continuous extension on $H^{1/2}$ and $\tilde{H}^{-1/2}(\Gamma)$ of the form $\langle f, v\rangle = \int\limits_\Gamma f\bar{v}\,dl$.

If W^h is a finite-dimensional space, the Galerkin scheme yields a system of linear algebraic equations. Let $\{\varphi_h^i\}$ be a basis in W^h. Then, (1.27) is equivalent to

$$D^h a_h = b_h, \tag{1.28}$$

where

$$D^h = \left\{\langle D\varphi_h^i, \varphi_h^j\rangle\right\}_{i,j}, \quad b_h = \left\{\langle f, \varphi_h^j\rangle\right\}_j^T,$$
$$a_h = \{a_h^i\}_i^T, \quad \varphi_h = \sum_i a_h^i \varphi_h^i.$$

Statement 3 *There exists $h_0 > 0$ such that for all $0 < h < h_0$, equation (1.27) has a unique solution $\varphi_h \in W^h$. For $h \to 0$, φ_h converge to the unique solution φ in the quasi-optimal sense; that is, there exists a constant C such that for all $0 < h < h_0$, the estimate*

$$\|\varphi - \varphi_h\|_{-1/2} \leq C \inf_{v \in W^h} \|\varphi - v\|_{-1/2} = C\tilde{E}_{-1/2}(\varphi, W^h) \tag{1.29}$$

is valid.

Estimate (1.29) is similar to (1.17), but here, $\|\cdot\|_{-1/2} = \|\cdot\|_{\tilde{H}^{-1/2}(\Gamma)}$.

Consider the equation

$$N\psi = -g.$$

Let $\{W^h\}_{h>0}$ be a family of subspaces of the space $\tilde{H}^{1/2}(\Gamma)$ such that for every $v \in \tilde{H}^{1/2}(\Gamma)$, $\left\|v - Q^h v\right\|_{1/2} \to 0$ for $h \to 0$, where Q^h is an orthogonal projector on W^h in $\tilde{H}^{1/2}(\Gamma)$. The Galerkin method consists in finding $\psi_h \in W^h$ such that

$$\langle N\psi_h, v\rangle = -\langle g, v\rangle, \quad \forall v \in W^h. \tag{1.30}$$

Here, the brackets denote the antiduality property of spaces $H^{-1/2}$ and $\tilde{H}^{1/2}(\Gamma)$, that is, a continuous extension on $H^{-1/2}$ and $\tilde{H}^{1/2}(\Gamma)$ of the form $\langle g, v\rangle = \int_\Gamma g\bar{v}\,dl$.

If W^h is a finite-dimensional space, the Galerkin scheme (1.19) yields a system of linear algebraic equations. Let $\{\psi_h^i\}$ be a basis in W^h. Then, (1.30) is equivalent to

$$N^h a_h = -b_h, \tag{1.31}$$

where

$$N^h = \left\{\langle N\psi_h^i, \psi_h^j\rangle\right\}_{i,j}, \quad b_h = \left\{\langle g, \psi_h^j\rangle\right\}_j^T,$$
$$a_h = \{a_h^i\}_i^T, \quad \psi_h = \sum_i a_h^i \psi_h^i.$$

Statement 4 *There exists $h_0 > 0$ such that for all $0 < h < h_0$, equation (1.30) has a unique solution $\psi_h \in W^h$. For $h \to 0$, ψ_h converge to the unique solution ψ in the quasi-optimal sense; that is, there exists a constant C such that for all $0 < h < h_0$, the estimate*

$$\|\psi - \psi_h\|_{1/2} \leq C \inf_{v \in W^h} \|\psi - v\|_{1/2} = C\tilde{E}_{1/2}(\psi, W^h) \tag{1.32}$$

is valid.

Estimate (1.32) is similar to (1.22), but here, $\|\cdot\|_{1/2} = \|\cdot\|_{\tilde{H}^{1/2}(\Gamma)}$.

Taking into account the asymptotic behavior (1.25) and (1.26) of the solutions to equations (1.10) and (1.11), it is easy to show that, according to Ref. [94], estimates (1.29) and (1.32) cannot provide the rate of convergence better than $O(h^{1/2-\varepsilon})$ when splines are taken as the basis functions of the Galerkin method. This is connected with a poor approximation by splines in the vicinities of endpoints of $\partial\Gamma$. The rate of convergence can be improved if to apply basis functions with a weight that takes into account (1.25) and (1.26). Consider a widely-distributed version of the Galerkin method for solving equations (1.10) and (1.11), in which the Chebyshev polynomials with weight [53] are applied as basis functions.

Numerical solution of equations of the first kind is based on the idea of regularization of a given equation. In one of the methods, the principal part of the equation is taken into account analytically. The principal part of operators D and N usually corresponds to a part of operators containing the singularity of the kernel of the integral equation. The regularization procedure consists of the following steps. First, the principal part of operators D and N is separated analytically, and then, the problem is solved numerically, so that in the numerical algorithm, the action of the principal part is taken into account explicitly. Sometimes, a different procedure is used: first, a numerical method is applied, and then, regularization is performed, for example, in the form of explicit summation of slowly convergent series or integrals, which yields the same reduced matrix equation.

Consider the Galerkin method for solving equations (1.10) and (1.11) in the case of one unclosed contour Γ_1 ($\Gamma = \Gamma_1$). Performing the change of variables, we reduce (1.10) and (1.11) to equations on the interval

$$\tilde{D}\tilde{\varphi} := \int_{-1}^{1} \left(\ln \frac{1}{|t-t_0|} + K_1(t,t_0) \right) \tilde{\varphi}(t)\, dt = \tilde{f}(t_0) \tag{1.33}$$

and

$$\tilde{N}\tilde{\psi} := \frac{\partial}{\partial t_0} \int_{-1}^{1} \frac{\partial}{\partial t} \left(\ln \frac{1}{|t-t_0|} + K_2(t,t_0) \right) \tilde{\psi}(t)\, dt = \tilde{g}(t_0), \tag{1.34}$$

$$t_0 \in (-1,1),$$

where singularities in the kernels are separated explicitly, and functions K_1 and K_2 are smooth. The variable t may be chosen as the natural parameter with the origin in the middle of the curve, normalized by $l_0/2$, where l_0 is the length of the curve. In the course of numerical solution of (1.33) and (1.34), it is expedient to separate at once the weight factor characterizing the behavior of solution in the vicinities of endpoints. We will look for approximate solutions in the form

$$\tilde{\varphi}_N(t) = \frac{1}{\sqrt{1-t^2}} \sum_{i=0}^{N-1} a_i^{(1)} T_i(t), \tag{1.35}$$

for equation (1.33), and

$$\tilde{\psi}_N(t) = \sqrt{1-t^2} \sum_{i=0}^{N-1} a_i^{(2)} U_i(t), \tag{1.36}$$

for equation (1.34), where $T_i(t)$ and $U_i(t)$ are the Chebyshev polynomials of the first and the second kind. Such a representation enables us to calculate the action of the principal parts of operators on functions $\tilde{\varphi}_N$ and $\tilde{\psi}_N$ in (1.33) and (1.34) explicitly:

$$\int_{-1}^{1} \ln \frac{1}{|t-t_0|} \tilde{\varphi}_N(t)\, dt = \pi a_0^{(1)} \ln 2 + \pi \sum_{i=1}^{N-1} \frac{1}{i} a_i^{(1)} T_i(t_0), \tag{1.37}$$

and

$$\frac{\partial}{\partial t_0} \int_{-1}^{1} \frac{\partial}{\partial t} \ln \frac{1}{|t-t_0|} \tilde{\psi}_N(t)\, dt = \pi \sum_{i=0}^{N-1} (i+1) a_i^{(2)} U_i(t_0), \tag{1.38}$$

We take the Chebyshev polynomials of the first and the second kind as probe functions for equations (1.33) and (1.34), respectively. Then, unknown coefficients $a_i^{(1)}$ and $a_i^{(2)}$ in (1.35) and (1.36) are determined from the systems of linear algebraic equations

$$\left(\tilde{D}\tilde{\varphi}_N, T_j \right)_{(1)} = \left(\tilde{f}, T_j \right)_{(1)}, j = 0, \ldots, N-1, \tag{1.39}$$

and

$$\left(\tilde{N}\tilde{\psi}_N, U_j \right)_{(2)} = \left(\tilde{g}, U_j \right)_{(2)}, j = 0, \ldots, N-1; \tag{1.40}$$

where $(\cdot,\cdot)_{(1)}$ and $(\cdot,\cdot)_{(2)}$ denote the inner products in the weighted spaces L_2:

$$(u,v)_{(1)} = \int_{-1}^{1} u(t)\overline{v}(t)\frac{1}{\sqrt{1-t^2}}\,dt,$$

$$(u,v)_{(2)} = \int_{-1}^{1} u(t)\overline{v}(t)\sqrt{1-t^2}\,dt.$$

By virtue of (1.37) and (1.38), the principal parts of operators $\tilde{D}$ and $\tilde{N}$ yield diagonal matrices

$$diag(\pi^2 \ln 2, \frac{\pi^2}{2}, \ldots, \frac{\pi^2}{2}\frac{1}{N-1}), \tag{1.41}$$

and

$$diag(\frac{\pi^2}{2}, \ldots, \frac{\pi^2}{2}N), \tag{1.42}$$

Other integrals in (1.39) and (1.40) are calculated approximately with the help of quadrature formulas. These numerical methods are stable and demonstrate high efficiency for screens that have small dimensions. One can assume that a screen is small if its length Γ is small compared with the wavelength; that is, $kl_0 \ll 1$. In this case, functions φ and ψ are well approximated already by several Chebyshev polynomials (usually, it is sufficient to take 5–10 polynomials depending on the required calculation accuracy). Indeed, the Chebyshev polynomials are eigenfunctions of the principal parts of operators $\tilde{D}$ and $\tilde{N}$, which enables one to obtain diagonal principal parts (1.41) and (1.42) in the matrix equation. Therefore, taking into account, for small screens, the behavior of solution in the vicinities of endpoints by introducing weight factors is of major importance, because endpoints essentially define the solution on the screen (for large screens, this conclusion is not valid). Calculations show that the methods are sufficiently efficient already for $kl_0 \sim 1$.

It follows from estimates (1.29) and (1.30) that in the general case, the smoother the solutions φ and ψ, the higher the rate of convergence. Functions φ and ψ are smooth when right-hand sides f and g in (1.10) and (1.11) are smooth. As follows from the regularity theorems for elliptic equations, if $f \in H^s(\Gamma)$ and $g \in H^{s-1}(\Gamma)$, then $\varphi \in H^{s-1}_{loc}(\Gamma)$ and $\psi \in H^s_{loc}(\Gamma)$, $s \geq \frac{1}{2}$. Consequently, if $f, g \in C^\infty(\overline{\Gamma})$, then $\varphi, \psi \in C^\infty(\Gamma)$. However, these assertions do not give us information concerning the behavior of φ and ψ in the vicinities of endpoints of $\partial\Gamma$. Exact results describing the behavior of φ and ψ in the vicinity of $\partial\Gamma$ with respect to smoothness of f and g (and smoothness of Γ) are obtained by the methods of the PDO theory on manifolds with an edge (in particular, on rectilinear beams [67]). Below, we will analyze these problems considering, as an example, the solution to the Dirichlet problem by the Galerkin method with splines φ_h^i and ψ_h^i taken as the basis functions. As far as the rate of convergence is concerned of the Galerkin methods considered above, in which the Chebyshev polynomials are applied, one can prove that if $f, g \in C^\infty(\overline{\Gamma})$, then

$$\|\varphi - \varphi_h\|_{-1/2} = O(h^p) \tag{1.43}$$

and

$$\|\psi - \psi_h\|_{1/2} = O(h^p) \tag{1.44}$$

as $h = \frac{1}{N} \to 0$ for every $p > 0$ (so that $\|\varphi - \varphi_h\|_{-1/2} \to 0$ and $\|\psi - \psi_h\|_{1/2} \to 0$ faster than an arbitrary power of h when $h \to 0$).

Consider application of the Galerkin method with splines φ_h^i taken as the basis functions to solving the equation $D\varphi = f$. Assume that $\overline{\Gamma}$ is parametrized so that $t \in [0,1]$, where t is the natural parameter, normalized by the length l_0 of curve Γ. Let $\{t_i\}_{i=0}^N$ be the uniform partition of interval $[0,1]$ with $t_0 = 0$, $t_N = 1$, and the step $h = \frac{1}{N}$. By $S_h^{r,k}(\Gamma)$, we denote the space of regular (r,k)-systems [95] (splines on Γ) formed by functions $\mu(t)$, which are splines with respect to variable t, where t is identified with the corresponding point on Γ, so that one may assume that $\mu(t)$ is defined on Γ. Parameters r and k are defined as follows: function $\mu(t)$ is a spline of order $r-1$ with respect to t, and parameter k provides the validity of the inclusion

$$\mu \in S_h^{r,k}(\Gamma) \subset \tilde{H}^k(\Gamma). \tag{1.45}$$

This inclusion imposes a restriction on the spline defect. For example, in the spaces $S_h^{2,1}$ and $S_h^{3,2}$, which we use in our subsequent considerations, the only possibility is that the spline defect equals unity, because, according to the embedding theorem [57], functions that belong to $\tilde{H}^1(\Gamma)$ and $\tilde{H}^2(\Gamma)$ are, respectively, continuous and continuously differentiable.

We define finite-dimensional spaces of regular splines on Γ with the partition $\{t_j\}$ as

$$\tilde{H}_h^1(\Gamma) := \{\varphi_0 = \sum_j \gamma_j \mu_j(t)|_\Gamma : \mu_j \in S_h^{2,1}(\Gamma),\ \varphi_0(0) = \varphi_0(1) = 0\}$$

and

$$\begin{aligned}\tilde{H}_h^2(\Gamma) \quad := \quad & \Big\{\varphi_1 = \sum_j \gamma_j \mu_j(t)|_\Gamma : \mu_j \in S_h^{3,2}(\Gamma), \\ & \varphi_1(0) = \varphi_1(1) = \varphi_1'(0) = \varphi_1'(1) = 0\Big\}.\end{aligned}$$

By virtue of (1.45), we have

$$\tilde{H}_h^1(\Gamma) \subset \tilde{H}^1(\Gamma) \subset \tilde{H}^{-1/2}(\Gamma),$$

$$\tilde{H}_h^2(\Gamma) \subset \tilde{H}^2(\Gamma) \subset \tilde{H}^{-1/2}(\Gamma).$$

Assume that $\chi(t) \in C^\infty$ is a cut function such that $0 \le \chi \le 1$ and

$$\chi(t) = \begin{cases} 1, & 0 \le t \le h/3,\ 1 - h/3 \le t \le 1 \\ 0, & h/2 \le t \le 1 - h/2 \end{cases}$$

and $\chi = 1$ in the vicinity of $\partial\Gamma$ as a function defined on Γ. Define the weighted spaces of splines

$$Z_h^{1/2}(\Gamma) = \{\varphi = \varphi_0 + \sum_{i=1}^{2} \alpha_i \rho_i^{-1/2} \chi : \varphi_0 \in \tilde{H}_h^1(\Gamma), \alpha_i \in \mathrm{R}\}$$

and

$$Z_h^{3/2}(\Gamma) = \left\{\varphi = \varphi_1 + \sum_{i=1}^{2} \left(\alpha_i \rho_i^{-1/2} + \beta_i \rho_i^{1/2}\right) \chi,\ \varphi_1 \in \tilde{H}_h^2(\Gamma),\ \alpha_i, \beta_i \in \mathrm{R}\right\},$$

where ρ_i is the distance to the endpoints of $\partial\Gamma$. These spaces are subspaces of $\tilde{H}^{-1/2}(\Gamma)$:

$$Z_h^{1/2}(\Gamma) \subset \tilde{H}^{-1/2}(\Gamma), \quad Z_h^{3/2}(\Gamma) \subset \tilde{H}^{-1/2}(\Gamma).$$

The Galerkin method for solving equation (1.10) with the choice of subspaces $Z_h^{1/2}(\Gamma)$ and $Z_h^{3/2}(\Gamma)$ consists in finding $\varphi_h \in Z_h^l(\Gamma)$ such that

$$\langle D\varphi_h, v\rangle = \langle f, v\rangle, \quad \forall v \in Z_h^l(\Gamma), \quad l = \frac{1}{2}, \frac{3}{2}. \tag{1.46}$$

The Galerkin scheme (1.46) yields a system of linear algebraic equations with respect to unknown coefficients γ_i, α_i, and β_i.

Statement 5 *There exists $h_0 > 0$ such that for all $0 < h < h_0$, equation (1.46) has a unique solution $\varphi_h \in Z_h^l(\Gamma)$. For $h \to 0$, φ_h converge to the unique solution φ. There exists a constant C such that for all $0 < h < h_0$, the estimate*

$$\|\varphi - \varphi_h\|_{-1/2} \leq Ch^{r+1/2} \|\varphi\|_{Z^r(\Gamma)} \tag{1.47}$$

is valid under the condition that

$$f \in H^{r+1}(\Gamma), \ -\frac{1}{2} \leq r \leq l + \sigma, \quad |\sigma| < \frac{1}{2}, \quad l = \frac{1}{2}, \frac{3}{2}.$$

If function f is sufficiently smooth, it follows from estimate (1.47) that when the behavior of solutions in the vicinities of endpoints of $\partial\Gamma$ is taken into account in a more accurate manner, the order of the rate of convergence increases. The results obtained with the help of estimate (1.47) are listed in the table:

The Galerkin method	$l = -1/2$	$l = 1/2$	$l = 3/2$
Order of the rate of convergence	$1/2 - \varepsilon$	$3/2 - \varepsilon$	$5/2 - \varepsilon$

Here, parameter l corresponds to the choice of the Galerkin method: $l = -1/2$, when the singularity in the vicinities of $\partial\Gamma$ is not taken into account [95]; $l = 1/2$, when the space $Z_h^{1/2}(\Gamma)$ is chosen, so that the singularity is taken into account in the form

$$\sum_{i=1}^{2} \alpha_i \rho_i^{-1/2} \chi;$$

and $l = 3/2$, when the singularity is taken into account in the form

$$\sum_{i=1}^{2} \left(\alpha_i \rho_i^{-1/2} + \beta_i \rho_i^{1/2}\right) \chi,$$

which corresponds to the choice of $Z_h^{3/2}(\Gamma)$. The rate of convergence is estimated as $\|\varphi - \varphi_h\|_{-1/2} = O(h^s)$ for $h \to 0$. The order of s is given in the table and $\varepsilon > 0$ is an arbitrarily small constant. Note that in other (nonequivalent) norms, the order of the rate of convergence may be different (see Ref. [95]); however, in all cases, the rate of convergence is improved when the behavior of solution in the vicinities of endpoints is taken into account more accurately.

where r_∂ is the distance to the endpoints of $\partial\Gamma$. These spaces are subspaces of [illegible]

$$[illegible]$$

The Galerkin method for solving equation (1.1) with the choice of subspace [illegible] consists in finding $\sigma_h \in$ [illegible] such that

$$[illegible] \qquad (1.45)$$

The relations (1.45) yield a system of linear algebraic equations with respect to the unknown coefficients α_j and β_j.

[illegible] for $0 < \varepsilon \leq 1$ [illegible] equation (1.45) has a unique solution [illegible] corresponding to the [illegible] and the following estimate holds:

$$[illegible] \qquad (1.46)$$

In addition, we have that

$$[illegible]$$

If function f is sufficiently smooth, it follows from [illegible] the solution [illegible] the order of the rate of convergence [illegible]. The respective [illegible] are listed in the table.

Galerkin method	[illegible]
Order of the rate of convergence	[illegible]

If the parameter [illegible] is chosen [illegible] singularity [illegible] the solution [illegible] when the space [illegible] is chosen [illegible] singularity [illegible] takes into account the form

$$[illegible]$$

and [illegible] the singularity [illegible]

$$[illegible]$$

which corresponds to the choice of [illegible]. The rate of convergence [illegible] for [illegible]. The [illegible] given in the table [illegible] constant [illegible] the order of the rate of convergence may be [illegible]. In all cases the rate of convergence is improved when the behavior of solution in the vicinity of [illegible] is taken into account more [illegible].

Chapter 2

Diffraction by a bounded planar screen

In this chapter, we consider the vector problem of diffraction of an electromagnetic field by a bounded, planar perfectly conducting screen with a piece-wise smooth boundary. We devote a separate chapter to the case of a planar screen by two reasons. First, this case is not elementary and is of independent practical interest; in addition to this, planar screens have not been investigated until recently. Second, the problem of diffraction by a planar screen is a fundamental problem, since here, all peculiarities of the general problem of diffraction by an arbitrary bounded screen are revealed in a sufficiently complete manner, as well as of the method of solution. At the same time, only in this case one can describe explicitly the spaces of solutions and images and the corresponding PDOs in terms of the Fourier transform and to apply a procedure of calculating the symbols of operators which does not require consideration of PDOs, acting in cross sections of vector bundles, as operators acting on manifolds. Third, a number of important results obtained for the problem of diffraction by a planar screen are not valid in the general case (for example, a diagonal splitting of a PDO considered on subspaces W_1 and W_2, positive definiteness of the corresponding quadratic forms, and the proof of the PDO injectivity, in which the uniqueness theorem is not used).

In Section 2.1, we consider a quasi-classical statement of the diffraction problem. The setting of this problem differs from conventional formulations by the fact that we do not specify the behavior of solutions in the vicinities of edges and corner points of a screen and impose a more general condition that the scattered field belongs to the space $L^2_{loc}(\mathrm{R}^3)$ (that is, energy is finite in every bounded spatial domain). We use this approach because, in our subsequent analysis, we will consider generalized solutions to the integrodifferential equation on the screen, and imposing additional edge conditions will excessively narrow the space of solutions. At the same time, it is not necessary to consider generalized solutions in all space R^3, because one can easily prove that the scattered field will be smooth everywhere outside the screen and continuous up to the screen surface from both sides except for the boundary points. Such a setting of the problem enables us to avoid unnecessary complications connected with generalized solutions, when vector potentials are constructed and analyzed.

Then, we prove the uniqueness theorem for the diffraction problem using the energy identities that follow from the Lorentz lemma. Note that this lemma is usually employed for smooth fields or for the fields that have a known behavior in the vicinities of edges of a screen, while, in the case under study, the only information available concerning the behavior of fields near edges is that they belong to the space $L^2_{loc}(\mathrm{R}^3)$. In Section 2.1, we extend the results to this case (which is not elementary).

In Section 2.2, we introduce the vector spaces of distributions defined on W and W'; the integrodifferential equation on the screen will be considered on these spaces. We prove statements that describe basic properties of these spaces; the most important of them concerns an expansion of W into a direct sum of orthogonal subspaces W_1 and W_2 (or W^1 and W^2 for W'), which enables us to obtain a diagonal splitting of the PDO and to investigate its properties.

In Section 2.3, we study representations of fields in the form of vector potentials and derive the main integrodifferential equation on the screen. The solution to the diffraction problem obtained with the help of vector potentials is not only natural and convenient from the theoretical viewpoint, but is also important for various applications, since this very method is usually applied for practical calculations.

In Section 2.4, the integrodifferential equation is considered as a pseudodifferential equation. The main (formally) matrix symbol of the latter equation turns out to be degenerate, which hampers substantially its analysis on a Cartesian product of two pieces of a certain space. We overcome this difficulty by considering the equation on an asymmetric space W coordinated with the quadratic form of the PDO. We define the generalized solution $u \in W$ of the pseudodifferential equation.

Section 2.5 is a central part of Chapter 2. Here, we perform a diagonal splitting of the PDO on subspaces W_1 and W_2, which is a key point in the analysis of its properties. Considering the restriction of PDOs on W_1 and W_2, we reveal the structure of the complete symbol of the operator and prove the Fredholm property of the operator with the zero index acting on the pair of spaces $W \to W'$. Significance of these results exceeds the bounds of the solvability theory developed in this chapter. Knowledge of the structure of a PDO is very important when one has to perform a correct choice of an appropriate numerical method for solving a pseudodifferential equation and basis and probe functions in the Galerkin method, when convergence of a numerical algorithm is analyzed, and so on.

One more important application of the results obtained in Section 2.5 concerning the smoothness of generalized solutions to pseudodifferential equations with smooth right-hand sides, is considered in Section 2.6. Here, among a variety of important practical problems, we note elucidation of the order of singularity of a solution to a pseudodifferential equation in the vicinities of the boundary and its corner points. Reducing the general vector pseudodifferential equation to two equations of the form $(1-\Delta)^{\mp 1/2} = f$ and using the properties of solutions to these equations in the Sobolev spaces [88], we establish, for the vector problem, the order of singularity of a solution in the vicinities of the boundary points. At the end of the section, we prove the solvability of the initial vector diffraction problem with an arbitrary right-hand side and, as a consequence, the possibility of representing every solution in the form of a vector potential.

2.1 Statement of the diffraction problem and the uniqueness theorem

Let $\Omega \subset \mathrm{R}^2 = \{x_3 = 0\} \subset \mathrm{R}^3$ be a bounded domain with a piece-wise smooth boundary Γ formed by a finite number of simple C^∞-arcs that join at nonzero angles. The problem of the diffraction of the monochromatic electromagnetic field E^0, H^0 by an infinitely thin, perfectly conducting screen Ω situated in the free space with the wavenumber k, $k^2 = \omega^2\mu(\varepsilon + i\sigma\omega^{-1})$, $\Im k \geq 0$ $(k \neq 0)$, is reduced to determination of the scattered electromagnetic field

$$E, H \in C^2(\mathrm{R}^3 \setminus \overline{\Omega}) \bigcap_{\delta>0} C(\overline{\mathrm{R}^3_+} \setminus \Gamma_\delta) \bigcap_{\delta>0} C(\overline{\mathrm{R}^3_-} \setminus \Gamma_\delta) \tag{2.1}$$

satisfying the homogeneous Maxwell equations

$$\begin{aligned} Rot H &= -ikE, \\ Rot E &= ikH, \quad \mathrm{x} \in \mathrm{R}^3 \setminus \overline{\Omega}, \end{aligned} \tag{2.2}$$

the boundary conditions for the tangential components of the electric field on the screen surface

$$E_\tau|_\Omega = -E^0_\tau|_\Omega, \tag{2.3}$$

the condition

$$E, H \in L^2_{loc}(\mathrm{R}^3) \tag{2.4}$$

that provides finiteness of energy in every bounded spatial domain, and the conditions at infinity

$$E, H = o(r^{-1}), \quad r := |\mathrm{x}| \to \infty \text{ for } \Im k > 0; \tag{2.5}$$

$$H \times e_r - E = o(r^{-1}), \quad E \times e_r + H = o(r^{-1}),$$

$$E, H = O(r^{-1}), \quad r \to \infty \text{ for } \Im k = 0 \tag{2.6}$$

(the Silver–Müller condition [27]). Here, $e_r = \mathrm{x}/|\mathrm{x}|$, $\times$ denotes the vector product, and $\Gamma_\delta := \{\mathrm{x} : |\mathrm{x} - y| < \delta, y \in \Gamma\}$. Electromagnetic fields are time-harmonic (the factor $exp(-i\omega t)$ is omitted), $\omega > 0$ is the circular frequency, $\varepsilon > 0$ and $\mu > 0$ are the permittivity and permeability, and $\sigma \geq 0$ is the conductivity of the medium. For the total field, we have $E^{\mathrm{tot}} = E^0 + E$, $H^{\mathrm{tot}} = H^0 + H$.

We will assume that all sources of the incident field are situated outside the screen $\overline{\Omega}$, so that, for a certain $\delta > 0$,

$$E^0 \in C^\infty(\Omega_\delta), \quad \Omega_\delta = \{\mathrm{x} : |\mathrm{x} - y| < \delta, y \in \Omega\} \tag{2.7}$$

and, consequently,

$$E^0_\tau|_\Omega \in C^\infty(\overline{\Omega}) \tag{2.8}$$

Usually, the incident field is either a plane wave or the electric or magnetic dipole [2] situated outside $\overline{\Omega}$. In these cases conditions (2.7) and (2.8) are fulfilled. The field E^0, H^0 is a solution to the system of Maxwell equations in the free space without the screen

Definition 1 *The solution E, H of problem (2.2)–(2.6) that satisfies condition (2.1) will be called quasi-classical.*

Such a name of the solution is caused by the fact that, first, in the classical formulation, one looks for a smooth solution continuous up to Ω (from each side); second, in (2.1)–(2.6), the behavior of the solution in the vicinity of Γ is not specified and general condition (2.4) is imposed (the solution of the problem will not be continuous up to $\overline{\Omega}$, and, in the vicinity of Γ, functions E and H have singularities). Often, condition (2.4) is replaced by the stronger Meixner condition [33], specifying the order of singularity of the field components in the vicinities of an edge. We note in this way that, in the vicinities of the boundary points of Γ, such conditions are not known (they will be discussed in Section 2.6).

Conditions at infinity (2.6) are equivalent to the Sommerfeld conditions for $\Im k = 0$ and $k \neq 0$

$$\frac{\partial}{\partial r}\begin{pmatrix} E \\ H \end{pmatrix} - ik\begin{pmatrix} E \\ H \end{pmatrix} = o\left(r^{-1}\right), \quad \begin{pmatrix} E \\ H \end{pmatrix} = O\left(r^{-1}\right), \quad r \to \infty, \tag{2.9}$$

and it is sometimes easier to verify them. The corresponding assertions are proved in Ref. [27]. Conditions (2.5), (2.6), and (2.9) hold uniformly along all directions e_r.

We will prove the uniqueness theorem for problem (2.1)–(2.6) using the energy identity obtained with the help of the Lorentz lemma. This lemma is usually employed for smooth fields E and H [23], while in the case under consideration, E and H have singularities in the vicinities of Γ. Generalization of the known results to this case, which is not elementary, is given below. Note that when the uniqueness theorem is proved, we do not use the fact that the screen is planar; therefore, in the case of arbitrary screens considered in Chapter 3, we omit a proof of a similar statement.

Let $G \subset \mathrm{R}^3$ be a bounded domain with the smooth boundary ∂G containing $\overline{\Omega}$: $\overline{\Omega} \subset \partial G$. By virtue of (2.2) and (2.4), fields E and H belong to the space

$$L^2_{Rot}(G) := \{F \in L_2(G) : Rot\, F \in L_2(G)\}$$

with the Hilbert norm

$$\|F\|^2_{Rot} = \|F\|^2_{L_2(G)} + \|Rot\, F\|^2_{L_2(G)}.$$

When the domain of definition should be specified, we will write $\|F\|_{Rot,G}$. Introducing the bilinear form

$$\Phi(F,V) := \int\limits_G (V\, Rot\, F - F\, Rot\, V)\, dx$$

and applying the Cauchy–Bunyakovskii inequality, we obtain the estimate

$$|\Phi(F,V)| \leq \|F\|_{Rot}\|V\|_{Rot}.$$

It is easy to show that, under the condition

$$\|F - F_n\|_{Rot} \to 0, \quad \|V - V_m\|_{Rot} \to 0,$$
$$F_n, V_m \in L^2_{Rot}(G), \quad n, m \to \infty,$$

the latter estimate yields

$$\Phi(F,V) = \lim_{n,m\to\infty} \Phi(F_n, V_m).$$

Thus, the bilinear form $\Phi(F,V)$ is continuous on $L^2_{Rot}(G)$.

For smooth $F,\, V \in C^1(\overline{G})$, we may perform integration by parts

$$\Phi(F,V) = \langle \nu \times f, v\rangle, \quad \langle \nu \times f, v\rangle := \int\limits_{\partial G} (\nu \times f) v\, ds,$$

where $\nu(x)$ is the external normal to G at the point $x \in \partial G$ and f and v on the right-hand side correspond to the traces of functions F and V on ∂G. Since

$$\begin{aligned} |\langle \nu \times f, v\rangle| &\leq \|F\|_{Rot}\|V\|_{Rot} \\ &\leq \|F\|_{Rot}\|V\|_{H^1(G)} \leq C\|F\|_{Rot}\|v\|_{H^{1/2}(\partial G)}, \end{aligned}$$

$\langle \nu \times f, v \rangle$ defines a linear continuous functional on the Sobolev space $H^{1/2}(\partial G)$; then, according to Ref. [32], $\nu \times f \in H^{-1/2}(\partial G)$ and C may be taken as the norm of a linear bounded "raising" operator $\chi : H^{1/2}(\partial G) \to H^1(G)$. Note that for $F = E$ or $F = H$, $\nu \times f$ may be considered as usual functions $\nu \times E|_{\partial G}$ or $\nu \times H|_{\partial G}$ defined on $\partial G \setminus \Gamma$, since all finite measures on ∂G belong to $H^s(\partial G)$, $s < -1$ [57] (if $u \in H^{-1/2}(\partial G)$ and $supp\, u \subset \Gamma$, then $u = 0$).

Consider the smoothing of function $F \in L^2_{Rot}(G)$ in the vicinity of ∂G. It will be sufficient to smooth the function in the vicinity of the line of singularities. If function F is defined in a wider domain, then we will assume that the inclusion $F \in L^2_{Rot}(G)$ (and similar designations) denotes the restriction of a function on G.

Denote by γ a piece-wise smooth curve ∂G belonging to $\gamma \subset \partial G$.

Lemma 1 *Assume that* $F \in L^2_{Rot}(G) \bigcap C^2(G) \bigcap_{\delta>0} C(\overline{G} \setminus \gamma_\delta)$. *Then, for all* $\varepsilon > 0$ *and* $\delta > 0$, *there exists a function* $F_1 \in C^2(G) \bigcap C(\overline{G})$ *such that* $\|F - F_1\|_{Rot} < \varepsilon$ *and* $F_1 \equiv F$ *in* $\overline{G} \setminus \gamma_\delta$.

□ Choose a finite system of points on curve γ, $x_j \in \gamma$, so that for a certain $\delta_0 < \delta$ and sufficiently small (positive) numbers α_j, semivicinities $U_j \equiv \overline{G} \bigcap B_j$, where $B_j = \{\mathrm{x} : |\mathrm{x} - x_j| < \delta_0\}$, would cover γ, $\gamma \subset \bigcup_j U_j$, and a shift of U_j by a vector $\alpha_j \nu(x_j)$ would satisfy the conditions

$$U_j^1 := (U_j + \alpha_j \nu(x_j)) \bigcap \overline{G} \subset \gamma_\delta \bigcap \overline{G}$$

and

$$(U_j \bigcap \partial G + \alpha_j \nu(x_j)) \bigcap \overline{G} = \emptyset.$$

which means that when the part of the boundary $U_j \bigcap \partial G$ is shifted by $\alpha_j \nu(x_j)$, it finds itself outside $\overline{G}$, and the remaining part of U_j^1 lies in γ_δ. Obviously, there always exist such a δ_0 and points x_j, because ∂G is smooth.

Let us form the cutting finite functions $\zeta_j(\mathrm{x}) \in C^\infty(\mathrm{R}^3)$ that satisfy the conditions $supp\, \zeta_j \subset B_j$ and $\zeta(\mathrm{x}) = \sum_j \zeta_j(\mathrm{x}) \equiv 1$ when $dist\,(\mathrm{x}, \gamma) < \delta'$ for a certain $\delta' < \delta_0$, with $\gamma_{\delta'} \bigcap \overline{G} \subset \bigcup_j U_j$. Consider the function $F = \zeta F + (1 - \zeta) F$. The equality

$$Rot(\zeta_j F) = \zeta_j Rot\, F + (Grad\, \zeta_j) \times F$$

yields $\zeta_j F$, $(1 - \zeta) F \in L^2_{Rot}(G)$. By virtue of the continuity of functions from L_2 with respect to the L_2-norm [34], for every $\varepsilon_j > 0$, there exists an $\alpha_j > 0$ such that

$$\|\zeta_j F - F_j\|_{Rot} < \varepsilon_j, \quad F_j(\mathrm{x}) = \zeta_j(\mathrm{x} - \alpha_j \nu(x_j)) F(\mathrm{x} - \alpha_j \nu(x_j)),$$

and $F_j \in C^2(G) \bigcap C(\overline{G})$. Now it is sufficient to choose such small ε_j that $\sum_j \varepsilon_j < \varepsilon$, and to set $F_1 = (1 - \zeta) F + \sum_j F_j$. Then, $\|F - F_1\|_{Rot} = \|\sum_j (F_j - \zeta_j F)\|_{Rot} < \varepsilon$ and $F_1 \equiv F$ in $\overline{G} \setminus \gamma_\delta$. □

Let $B_R = \{\mathrm{x} : |\mathrm{x}| < R\}$ be a ball containing $\overline{G}$, $\overline{G} \subset B_R$, and $[\nu \times F]_{\partial G}$ $(\in H^{-1/2}(\partial G))$ be a difference of limiting values of $\nu \times F$ taken from different sides of ∂G. We will also introduce the notation $Q = B_R \setminus \overline{G}$.

Lemma 2 *Assume that*

$$F \in L^2_{Rot}(G) \bigcap C^2(G) \bigcap_{\delta>0} C(\overline{G} \setminus \Gamma_\delta),$$
$$F \in L^2_{Rot}(Q) \bigcap C^2(Q) \bigcap_{\delta>0} C(\overline{Q} \setminus \Gamma_\delta),$$

and $[\nu \times F]_{\partial G} = 0$. *Then,* $F \in L^2_{Rot}(B_R)$ *and for every* $\varepsilon > 0$ *and every* $\delta > 0$, *there exists a function* F_2 *satisfying the conditions*

$$F_2 \in C^2(G)\bigcap C(\overline{G}),\ F_2 \in C^2(Q)\bigcap C(\overline{Q}),\ [\nu \times F_2]_{\partial G} = 0,$$

such that $\|F - F_2\|_{Rot,B_R} < \varepsilon$ *and* $F_2 \equiv F$ *in* $\overline{B}_R \setminus \Gamma_\delta$.

□ First of all, using the condition $[\nu \times F]_{\partial G} = 0$, we perform integration by parts to obtain

$$\int\limits_{B_R} (Rot\ F)V\,dx = \int\limits_{B_R} F\ Rot\ V\,dx$$

for every smooth function V, which is finite in B_R ($supp\,V \subset B_R$); consequently, $F \in L^2_{Rot}(B_R)$. For a given δ, we choose $\delta' < \delta_0 < \delta$ and cutting function $\zeta(x) \in C^\infty(R^3)$ so that $\zeta(x) \equiv 1$ in $\Gamma_{\delta'}$ and $\zeta(x) \equiv 0$ outside Γ_{δ_0}. We have $\zeta F,\ (1-\zeta)F \in L^2_{Rot}(B_R)$. Let $\omega_h(|x|)$ be the averaging kernel of the radius $h < (\delta - \delta_0)$ [34]. Consider the function

$$(\zeta F)_h(x) = \int\limits_{\Gamma_{\delta 0}} \zeta(y)F(y)\,\omega_h(|x-y|)\,dy$$

averaged over the domain Γ_{δ_0}, for which $(\zeta F)_h \equiv 0$ outside Γ_δ. Since

$$\|\zeta F - (\zeta F)_h\|^2_{Rot} = \int\limits_{R^3} (1-\hat{\omega}_h(\xi))^2(|\widehat{F\zeta}(\xi)|^2 + |\xi \times \widehat{F\zeta}(\xi)|^2)\,d\xi,$$

where $\widehat{F\zeta}(\xi)$ is the Fourier transform of the finite function $F(x)\zeta(x)$, then, for $h \to 0$, we obtain $\|\zeta F - (\zeta F)_h\|_{Rot} \to 0$. Choose such a small value $h > 0$, that for $F_2 = (1-\zeta)F + (\zeta F)_h$, the condition

$$\|F - F_2\|_{Rot} = \|\zeta F - (\zeta F)_h\|_{Rot} < \varepsilon.$$

is valid; then, it is clear that $F_2 \equiv F$ in $\overline{B}_R \setminus \Gamma_\delta$ and for this function F_2, all conditions of the lemma are fulfilled. □

Lemma 3 *Assume that*

$$F \in L^2_{Rot}(G)\bigcap C^2(G) \bigcap_{\delta>0} C(\overline{G} \setminus \Gamma_\delta),$$

$$F \in L^2_{Rot}(Q)\bigcap C^2(Q) \bigcap_{\delta>0} C(\overline{Q} \setminus \Gamma_\delta),$$

and $[\nu \times F]_{\partial G \setminus \overline{\Omega}} = 0$ *(where* $\Gamma = \partial\Omega$*). Then, for every* $\varepsilon > 0$ *and every* $\delta > 0$, *there exist* Ω', $\overline{\Omega'} \subset \Omega$, *and the function*

$$F_3 \in C^2(G)\bigcap C(\overline{G}),\ F_3 \in C^2(Q)\bigcap C(\overline{Q}),$$

such that

$$\|F - F_3\|_{Rot,G} < \varepsilon,\ \|F - F_3\|_{Rot,Q} < \varepsilon,\ [\nu \times F_3]_{\partial G \setminus \overline{\Omega'}} = 0$$

and $F_3 \equiv F$ *in* $\overline{B}_R \setminus \Gamma_\delta$.

□ It is known [36] that in the vicinity $U \subset \partial G$ of every point $x \in \partial G$, one can introduce isothermal (conformal) regular coordinates u_1, u_2 on the surface ∂G, in which the first quadratic form of the surface will have the form $g^2(u_1, u_2)((du_1)^2 + (du_2)^2)$, where $g^2(u_1, u_2) \geq C_0 > 0$ is a smooth function. We take a countable set of points $x_j \in \Gamma$, dense everywhere in Γ, and introduce the coordinates u_1^j, u_2^j in the vicinity $U_j \subset \partial G$. From a countable subcover of Γ by vicinities U_j, we choose a finite subcover (which exists because ∂G is compact in R^3); we will denote this subcover by $\{U_j\}$. Consider a vicinity of a point $x_j \in U_j$ and choose the third unit coordinate vector perpendicular to ∂G in the direction of the external normal $\nu(x)$ and the coordinate u_3^j, so that the element of the length should be equal to

$$ds^2 = g_j^2(u_1^j, u_2^j)\,((du_1^j)^2 + (du_2^j)^2 + (du_3^j)^2).$$

Thus, in a certain vicinity of the point x_j, we have constructed curvilinear orthogonal coordinates u_1^j, u_2^j, u_3^j. Without loss of generality, one may assume that this vicinity is a ball $B_j = \{x : |x - x_j| < \delta_0\}$ of the radius $\delta_0 < \delta$, and $\Gamma \subset \bigcup_j B_j$.

Let $\zeta_j(x) \in C^\infty(R^3)$ be the cutting finite functions obeying the following conditions: $supp\,\zeta_j \subset B_j$; $\zeta(x) = \sum_j \zeta_j(x) \equiv 1$, where $dist(x, \Gamma) < \delta'$ for a certain $\delta' < \delta_0$; and $\Gamma_{\delta'} \subset \bigcup_j B_j$. Consider the function

$$\begin{aligned} F_j(u^j) &= F_j(u_1^j, u_2^j, u_3^j) \\ &= (\zeta_j F)(u_1^j - t\beta_1^j, u_2^j - t\beta_2^j, u_3^j) \\ &= (\zeta_j F)(u^j - t\beta^j), \end{aligned}$$

where

$$\beta^j = (\beta_1^j,\, \beta_2^j,\, 0).$$

Choose β^j so that for $\Gamma_j \equiv \Gamma \bigcap B_j + t\beta^j$, the conditions $\Gamma_j \subset \Omega$, $\Gamma_j \bigcap \Gamma = \emptyset$ should be satisfied. We will assume that the value of t is sufficiently small, $B_j + t\beta^j \subset \Gamma_\delta$. In the curvilinear coordinates u_1, u_2, u_3, the expression for the curl takes the form

$$\begin{aligned} Rot\ A(u) &= g^{-1}\left(a_1\left(\frac{\partial A_3}{\partial u_2} - \frac{\partial A_2}{\partial u_3}\right)\right. \\ &\quad + a_2\left(\frac{\partial A_1}{\partial u_3} - \frac{\partial A_3}{\partial u_1}\right) + a_3\left.\left(\frac{\partial A_2}{\partial u_1} - \frac{\partial A_1}{\partial u_2}\right)\right) \\ &\quad + g^{-2}\left(a_1 A_3 \frac{\partial g}{\partial u_2} - a_2 A_3 \frac{\partial g}{\partial u_1} + a_3\left(A_2 \frac{\partial g}{\partial u_1} - A_1 \frac{\partial g}{\partial u_2}\right)\right). \end{aligned}$$

Here, a_1, a_2, a_3 are the unit coordinate vectors and index j is omitted, as well as in all subsequent formulas. We will prove that if $A(u) \in L^2_{Rot}(G)$, then $A(u - t\beta) \in L^2_{Rot}(G)$ for sufficiently small t and $\|A(u) - A(u - t\beta)\|_{Rot,G} \to 0$ for $t \to 0$. Indeed, since $g(u)$ is smooth, it is sufficient to show, using the expression for the curl, that if a scalar function $f \in L_2(D)$, $supp\, f \subset D$, and the unit vector $a = a(u)$ are continuous with respect to u in $\overline{D}$, $a \in C(\overline{D})$, and $\|a\| = 1$, then $f(u - t\beta)\, a(u - t\beta) \in L_2(D)$ and

$$\|f(u - t\beta)\, a(u - t\beta) - f(u)\, a(u)\|_{L_2(D)} \to 0$$

for $t \to 0$; here, $D = B_j \bigcap \overline{G}$. We note that, since $supp\, f(u) \subset D$, then $supp\, f(u - t\beta) \subset D$ for small t. The latter statements follow from the relationships

$$\begin{aligned} \int_D |f(u-t\beta)\,a(u-t\beta)|^2\,dv &= \int_D |f(u-t\beta)|^2\,dv \\ &= \int_D |f(u-t\beta)|^2\,g^3(u)\,du \\ &= \int_D |f(u)|^2\,g^3(u+t\beta)\,du \end{aligned}$$

and the estimate

$$\int_D |f(u-t\beta)\,a(u-t\beta) - f(u)\,a(u)|^2\,dv \le 2\|f(u-t\beta) - f(u)\|^2_{L_2(D)}$$
$$+2 \max_{u \in supp\, f} |a(u-t\beta) - a(u)|^2 \|f(u)\|^2_{L_2(D)}.$$

Then, $\|F_j - \zeta_j F\|_{Rot,G} \to 0$ as $t \to 0$. One can prove in a similar manner that $\|F_j - \zeta_j F\|_{Rot,Q} \to 0$ as $t \to 0$.

Let us apply Lemma 1 to smooth F_j in the vicinity of Γ_j. For every $\varepsilon_j > 0$, there exists

$$F_j^1 \in L^2_{Rot}(G) \bigcap C^2(G) \bigcap C(\overline{G}),$$
$$F_j^1 \in L^2_{Rot}(Q) \bigcap C^2(Q) \bigcap C(\overline{Q})$$

such that

$$\|F_j^1 - \zeta_j F\|_{Rot,G} < \varepsilon_j, \quad \|F_j^1 - \zeta_j F\|_{Rot,Q} < \varepsilon_j, \quad supp\, F_j^1 \subset \Gamma_\delta.$$

Since the lines of singularities Γ_j of the function F_j belong to Ω and $\Gamma_j \bigcap \Gamma = \emptyset$, there exists a domain Ω_0, $\overline{\Omega}_0 \subset \Omega$, with a piece-wise smooth boundary $\partial\Omega_0 \subset \Omega$, such that $[\nu \times F_j]_{\partial G \setminus \overline{\Omega}_0} = 0$. Then, according to Lemma 1, functions F_j^1 can be chosen so that $[\nu \times F_j^1]_{\partial G \setminus \overline{\Omega}'} = 0$ for a certain domain Ω' satisfying the conditions $\Omega_0 \subset \Omega' \subset \Omega$ and $\partial\Omega' \bigcap \partial\Omega = \emptyset$.

Now, it is easy to see that all requirements of the lemma will be fulfilled if to take $\sum_j \varepsilon_j < \varepsilon$ and set $F_3 = (1 - \zeta)F + \sum_j F_j^1$. □

Theorem 1 *For $\Im k \ge 0$, $k \ne 0$, problem (2.1)–(2.6) has not more than one solution.*

□ It is sufficient to prove that the homogeneous problem (for $E^0_\tau|_\Omega \equiv 0$) has only a trivial solution. Let G, Q, and B_R denote the same quantities. Then, we can formulate the following conjugation problem for E and H. Fields E and H satisfy conditions (2.1) and (2.4), the Maxwell equations (2.2) in G and Q, and the boundary conditions $E_\tau|_\Omega = 0$, $[E_\tau]_{\partial G \setminus \overline{\Omega}} = 0$, and $[H_\tau]_{\partial G \setminus \overline{\Omega}} = 0$. Fix $\delta > 0$ such that $\overline{\Gamma}_\delta \subset B_R$. By virtue of Lemma 3, for every $\varepsilon > 0$, there exists

$$H^\varepsilon \in C^2(G) \bigcap C(\overline{G})\,,\; H^\varepsilon \in C^2(Q) \bigcap C(\overline{Q})$$

such that

$$\|H - H^\varepsilon\|_{Rot,G} < \varepsilon, \|H - H^\varepsilon\|_{Rot,Q} < \varepsilon \text{ and } [\nu \times H^\varepsilon]_{\partial G \setminus \overline{\Omega}'} = 0$$

for a certain domain Ω', $\overline{\Omega}' \subset \Omega$; note that $H^\varepsilon \equiv H$ in $\overline{B}_R \backslash \Gamma_\delta$. Choose $\delta' < dist\,(\partial\Omega, \partial\Omega')$, $\delta' < \delta$ $(\delta' > 0)$. According to Lemma 2, for $\varepsilon > 0$, there exists

$$E^\varepsilon \in C^2(G) \bigcap C(\overline{G}), E^\varepsilon \in C^2(Q) \bigcap C(\overline{Q})$$

with $[\nu \times E^\varepsilon]_{\partial G} = 0$, such that

$$\|E - E^\varepsilon\|_{Rot,G} < \varepsilon, \; \|E - E^\varepsilon\|_{Rot,Q} < \varepsilon, \; E^\varepsilon \equiv E \text{ in } \overline{B}_R \setminus \Gamma_{\delta'}.$$

Hence, $[(\nu \times E^\varepsilon)\overline{H}^\varepsilon]_{\partial G} = 0$ because H^ε_τ has different limiting values only on $\overline{\Omega}'$, and $E^\varepsilon_\tau|_{\overline{\Omega}'} \equiv 0$. Then, applying integration by parts to functions E^ε and $\overline{H}^\varepsilon$ in domains G and Q and summing up the results, we obtain

$$\int_G (\overline{H}^\varepsilon Rot E^\varepsilon - E^\varepsilon Rot \overline{H}^\varepsilon)\, dx + \int_Q (\overline{H}^\varepsilon Rot E^\varepsilon - E^\varepsilon Rot \overline{H}^\varepsilon)\, dx$$

$$= \int_{\partial B_R} (e_r \times E^\varepsilon)\overline{H}^\varepsilon\, ds$$

(we assume that, in the surface integral, the values of E^ε and H^ε are taken on ∂B_R). Since $E^\varepsilon \equiv E, H^\varepsilon \equiv H$ on ∂B_R, the right-hand side of the equality does not depend on ε. Passing to the limit $\varepsilon \to 0$ on the left-hand side of the equality and separating the imaginary part, we obtain, taking into account the Maxwell equations and that $\|E - E^\varepsilon\|_{Rot} \to 0$ and $\|H - H^\varepsilon\|_{Rot} \to 0$ as $\varepsilon \to 0$, the relationship

$$\Re \int_{\partial B_R} (E \times \overline{H}) e_r\, ds + \Im k \int_{B_R} (\|E\|^2 + \|H\|^2)\, dx = 0 \qquad (2.10)$$

For $\Im k > 0$, from (2.5) and (2.10), we find that

$$\int_{\partial B_R} (\|E\|^2 + \|H\|^2)\, dx \;\to 0 \quad \text{for} \quad R \to \infty,$$

which yields $E \equiv 0, H \equiv 0$ in R^3.

For $\Im k = 0$, from (2.6) and (2.10), we have

$$\int_{\partial B_R} |H|^2\, ds \;\to\; 0 \quad \text{for} \quad R \to \infty.$$

Then, since the H-field components satisfy the homogeneous Helmholtz equation outside $\overline{G}$ and the Sommerfeld conditions (2.9), we can use the Röllich lemma [23, 45] to conclude that $H \equiv 0$ outside B_R and, consequently, $H \equiv 0$ in R^3 because H is an analytical function. Applying the same reasoning, we see that, by virtue of the first Maxwell equation (2.2), $E \equiv 0$ in R^3. □

We note that the proof of the theorem is valid also for a system of a finite number of nonintersecting screens. In this case, the domain G should be chosen so that its boundary ∂G contain all screens.

2.2 Vector spaces W and W'

In order to study the problem of diffraction on the screen Ω, we introduce a vector space of distributions W.

For every real s, we set, according to Ref. [58],

$$H^s(\Omega) := \{u|_\Omega \ : u \in H^s(\mathrm{R}^2)\}$$

and

$$\tilde{H}^s(\overline{\Omega}) := \{u \in H^s(\mathrm{R}^2) : supp\ u \subset \overline{\Omega}\}.$$

The inner product and the norm are defined for $H^s(\mathrm{R}^2)$ in a usual manner:

$$(u,v)_s = \int \langle\xi\rangle^{2s}\widehat{u}(\xi)\overline{\widehat{v}(\xi)}\, d\xi, \quad \|u\|_s^2 = (u,u)_s,$$

where

$$\langle\xi\rangle = (1+\|\xi\|^2)^{1/2}.$$

Here and below, we assume that the integral, for which the domain of integration is not specified, is taken over the whole space R^2. $\tilde{H}^s(\overline{\Omega})$ is a closed subspace of $H^s(\mathrm{R}^2)$ with the induced inner product and the norm. In the space $H^s(\Omega) = H^s(\mathrm{R}^2)/\tilde{H}^s(\overline{\Omega})$, we introduce the inner product and the norm of the factor-space. $H^{-s}(\Omega)$ and $\tilde{H}^s(\overline{\Omega})$ are antidual spaces for all $s \in R$, and $\tilde{H}^s(\overline{\Omega})$ can be formed as a closure of $C_0^\infty(\Omega)$ in $H^s(\mathrm{R}^2)$ [58].

We will be mainly interested in spaces of vector-functions; therefore, by u and v, we will denote the vectors $u = (u_1,u_2)^T$ and $v = (v_1,v_2)^T$, and so on. When we write $u \in H^s$, we assume that H^s is a Cartesian product of two pieces of the space H^s with the inner product and the norm

$$(u,v)_s = (u_1,v_1)_s + (u_2,v_2)_s = \int \langle\xi\rangle^{2s}\widehat{u}(\xi)\cdot\overline{\widehat{v}(\xi)}\, d\xi,$$

$$\|u\|_s^2 = \|u_1\|_s^2 + \|u_2\|_s^2 = \int \langle\xi\rangle^{2s}|\widehat{u}(\xi)|^2\, d\xi.$$

In the vector case, we will retain the same designations for spaces, because it is always clear which concrete space is considered.

We define the Hilbert spaces $W = W(\overline{\Omega})$ as a supplement of $C_0^\infty(\Omega)$ with respect to the norm

$$\|u\|_W^2 = \int \frac{1}{\langle\xi\rangle}|\widehat{u}(\xi)|^2\, d\xi + \int \frac{1}{\langle\xi\rangle}|\xi\cdot\widehat{u}(\xi)|^2\, d\xi$$

with the inner product

$$(u,v)_W = \int \frac{1}{\langle\xi\rangle}\widehat{u}(\xi)\cdot\overline{\widehat{v}(\xi)}\, d\xi + \int \frac{1}{\langle\xi\rangle}(\xi\cdot\widehat{u}(\xi)(\xi\cdot\overline{\widehat{v}(\xi)})\, d\xi,$$

where $\widehat{u}$ denotes the Fourier transform of the distribution u.

Statement 6 $W = \{u \in \tilde{H}^{-1/2}(\overline{\Omega}) \ : \ div\, u \in \tilde{H}^{-1/2}(\overline{\Omega})\}$.

□ Let $u \in \tilde{H}^{-1/2}(\overline{\Omega})$ and $div\, u \in \tilde{H}^{-1/2}(\overline{\Omega})$. Assume first that Ω is a star-like domain with respect to a certain point $x_0 \in \Omega$. Without loss of generality, one can set $x_0 = 0$ (otherwise, the origin may be shifted by x_0). Consider the distribution $u_\alpha(x) = u(\alpha x)$, $\alpha \geq 1$. It is clear that $supp\, u_\alpha(x) \subset \Omega$ for $\alpha > 1$. Consider also the convolution $u_{\alpha,h} = u_\alpha * \omega_h$, where $\omega_h(x) = h^{-2}\omega_1(|x|/h)$ is the averaging kernel of the radius h [4] with $h < dist(supp\, u_\alpha(x), \Gamma)$ $(\alpha > 1)$; $\omega_h(x)$ form a δ-like sequence for $h \to 0$. Then, $u_{\alpha,h} \in C_0^\infty(\Omega)$. Let us show that $\|u - u_{\alpha,h}\|_W \to 0$ for $\alpha \to 1$. Introduce the quantity

$$I = \int \frac{1}{\langle\xi\rangle}\left(|\widehat{u} - \widehat{u}_{\alpha,h}|^2 + |\xi \cdot (\widehat{u} - \widehat{u}_{\alpha,h})|^2\right) d\xi$$

$$= \int\limits_{|\xi|>R} + \int\limits_{|\xi|\leq R} = I_1 + I_2,$$

where

$$\widehat{u}_{\alpha,h} = \alpha^{-2}\, \widehat{u}(\alpha^{-1}\xi)\, \widehat{\omega}_h(\xi).$$

For I_1, the following estimate is valid:

$$I_1 \leq C \int\limits_{|\xi|>R/2} \frac{1}{\langle\xi\rangle}\left(|\widehat{u}|^2 + |\xi \cdot \widehat{u}|^2\right) d\xi\ ,$$

where C does not depend on α $(1 \leq \alpha \leq 2)$ and h. Then, there exists an R, which does not depend on α and h, such that $I_1 < \varepsilon/2$ for every $\varepsilon > 0$. For $f \in C^\infty(|\xi| \leq R)$, we have

$$\begin{aligned}
\left|f(\xi) - \frac{1}{\beta} f\left(\frac{\xi}{\alpha}\right) \widehat{\omega}_h(\xi)\right| &\leq \left|f(\xi) - \frac{1}{\beta} f(\xi)\right| + \left|\frac{1}{\beta} f(\xi) - \frac{1}{\beta} f(\xi)\widehat{\omega}_h(\xi)\right| \\
&\quad + \left|\frac{1}{\beta} f(\xi)\widehat{\omega}_h(\xi) - \frac{1}{\beta} f\left(\frac{\xi}{\alpha}\right) \widehat{\omega}_h(\xi)\right| \\
&\leq C_1(\alpha - 1) + C_2|1 - \widehat{\omega}_h(\xi)| \\
&\quad + \left|f(\xi) - f\left(\frac{\xi}{\alpha}\right)\right| \\
&\leq C_3(\alpha - 1) + C_4 h < \varepsilon/2
\end{aligned}$$

for $\alpha \to 1$ $(h \to 0)$; here, $\beta = \alpha$ or α^2. Then, $I_2 < \varepsilon/2$ for $\alpha \to 1$ and, finally, $I < \varepsilon$ for $\alpha \to 1$ $(h \to 0)$.

In the case of an arbitrary domain, we choose a finite smooth partition of unity over $\overline{\Omega}$ so that the sets $P_\nu = supp\, \psi_\nu \cap \overline{\Omega}$, $P_\nu \subset supp\,(\psi_\nu u)$, should be star-like with respect to certain points $x_{0,\nu} \in int\, P_\nu$, $1 = \sum_\nu \psi_\nu$. Repeating the above considerations for every $\psi_\nu u$, we complete the proof of the theorem. □

The norm in W can be represented as

$$\|u\|_W^2 = \|u\|_{-1/2}^2 + \|div\, u\|_{-1/2}^2.$$

We define W_1 and W_2 as subspaces of W:

$$W_1 := \{u \in W : \forall\xi \quad \xi_1\widehat{u}_1(\xi) + \xi_2\widehat{u}_2(\xi) = 0\},$$

$$W_2 := \{u \in W : \forall\xi \quad \xi_2\widehat{u}_1(\xi) - \xi_1\widehat{u}_2(\xi) = 0\}$$

Note that here, $\widehat{u}_i(\xi) \in C^\infty(\mathrm{R}^2)$ and, therefore, the definition makes sense.

Statement 7 *The space W is expanded into a direct sum of (closed) orthogonal subspaces W_1 and W_2:*

$$W = W_1 \oplus W_2.$$

□ Let $u \in W_1$ and $v \in W_2$. Then, $\widehat{u}(\xi) \cdot \widehat{v}(\xi) = 0$ and, consequently, $W_1 \perp W_2$. The fact that subspaces W_1 and W_2 are closed can be verified directly. If $\|u - u_n\|_W \to 0$ as $n \to \infty$, where $u \in W$ and $u_n \in W_1$, then

$$\int \frac{1}{\langle \xi \rangle} |\xi \cdot (\widehat{u} - \widehat{u}_n)|^2 \, d\xi = \int \frac{1}{\langle \xi \rangle} |\xi \cdot \widehat{u}|^2 \, d\xi \to 0$$

and, therefore, $\xi \cdot \widehat{u} = 0$, $u \in W_1$. One can prove in a similar manner that if $\|v - v_n\|_W \to 0$ as $n \to \infty$, where $v \in W$ and $v_n \in W_2$, then

$$\int \frac{1}{\langle \xi \rangle} |\widehat{v} - \widehat{v}_n|^2 \, d\xi + \int \frac{1}{\langle \xi \rangle} |\xi \cdot (\widehat{v} - \widehat{v}_n)|^2 \, d\xi$$

$$= \int \frac{1}{\langle \xi \rangle} \left(1 + \frac{1}{|\xi|^2}\right) |\xi \cdot (\widehat{v} - \widehat{v}_n)|^2 \, d\xi + \int \frac{1}{\langle \xi \rangle |\xi|^2} |\xi_2 \widehat{v}_1 - \xi_1 \widehat{v}_2|^2 \, d\xi \to 0,$$

which yields $\xi_2 \widehat{v}_1 - \xi_1 \widehat{v}_2 = 0$, $v \in W_2$. For smooth vectors $g \in C_0^\infty(\Omega)$, the expansion $g = u + v$ coincides with the known expansions into solenoid and potential components. Taking into account that the density belongs to $C_0^\infty(\Omega)$ in W and the inclusion $C_0^\infty(\Omega) \subset W_1 \oplus W_2 \subset W$ is valid (note that W_1 and W_2 are closed), we complete the proof of the theorem. □

Statement 8 *The embeddings*

$$\tilde{H}^{1/2}(\overline{\Omega}) \subset W \subset \tilde{H}^{-1/2}(\overline{\Omega})$$

are continuous, and the norms can be estimated as

$$\|u\|_{-1/2} \le \|u\|_W \le \|u\|_{1/2},$$

where

$$\|u\|_W = \|u\|_{-1/2} \quad \textit{for} \quad u \in W_1$$

and

$$\|u\|_W = \|u\|_{1/2} \quad \textit{for} \quad u \in W_2.$$

The proof is reduced to a simple verification of the last two equalities.

Statement 9 *$W_1 \subset \tilde{H}^{-1/2}(\overline{\Omega})$ is a subspace (closed with respect to the norm $\| \cdot \|_{-1/2}$). $W_2 \subset \tilde{H}^{1/2}(\overline{\Omega})$ is a subspace (closed with respect to the norm $\| \cdot \|_{1/2}$).*

□ Let $\|u - u_n\|_{-1/2} \to 0$, $u \in \tilde{H}^{-1/2}(\overline{\Omega})$, and $u_n \in W_1$ as $n \to \infty$. Then,

$$\int \frac{1}{\langle \xi \rangle} |\widehat{u}_n - \widehat{u}| \, d\xi = \int \frac{|\xi|^{-2}}{\langle \xi \rangle} |\xi' \cdot (\widehat{u}_n - \widehat{u})|^2 \, d\xi + \int \frac{|\xi|^{-2}}{\langle \xi \rangle} |\xi \cdot \widehat{u}|^2 \, d\xi \to 0,$$

where $\xi' = (\xi_2, -\xi_1)$ and, consequently, $\xi \cdot \widehat{u} = 0$, $u \in W_1$. Assuming that $\|v - v_n\|_{1/2} \to 0$, $v \in \tilde{H}^{1/2}(\overline{\Omega})$, and $v_n \in W_2$ as $n \to \infty$ and applying a similar reasoning, we obtain

$$\int \langle \xi \rangle |\widehat{v}_n - \widehat{v}|^2 \, d\xi = \int \frac{\langle \xi \rangle}{|\xi|^2} |\xi \cdot (\widehat{v}_n - \widehat{v})|^2 \, d\xi + \int \frac{\langle \xi \rangle}{|\xi|^2} |\xi' \cdot \widehat{v}|^2 \, d\xi \to 0,$$

so that $\widehat{v} \cdot \xi' = 0$, $v \in W_2$. Hence, we have proved that subspaces W_1 and W_2 are closed with respect to the norms $\|\cdot\|_{-1/2}$ and $\|\cdot\|_{1/2}$, respectively. □

For the subspace $W' := (W(\overline{\Omega}))'$,

$$W' = \left\{ f|_{\Omega} \; : \; f \in H^{-1/2}(\mathrm{R}^2), \; \frac{\partial f_1}{\partial x_2} - \frac{\partial f_2}{\partial x_1} \in H^{-1/2}(\mathrm{R}^2) \right\},$$

which is antidual with respect to W, the following statement holds.

Statement 10 *The embeddings*

$$H^{1/2}(\Omega) \subset W' \subset H^{-1/2}(\Omega).$$

are continuous.

□ Denote by $'W(\mathrm{R}^2)$ the space of elements from $H^{-1/2}(\mathrm{R}^2)$ that have the finite norm

$$\|f\|'^2 := \|f\|^2_{-1/2} + \left\| \frac{\partial f_2}{\partial x_1} - \frac{\partial f_1}{\partial x_2} \right\|^2_{-1/2}.$$

According to the Riesz theorem, the general form of the antilinear continuous functional on $W(\mathrm{R}^2)$ is $(h,u)_W$ for a certain $h \in W(\mathrm{R}^2)$. The norm of the functional equals $\|h\|_W$, so that $(h,u)_W = \int \widehat{f} \cdot \overline{\widehat{u}}\, d\xi$, where $\widehat{f}(\xi) = \langle \xi \rangle^{-1}(\widehat{h} + \xi(\xi \cdot \widehat{h}))$. Using the identity

$$(\xi \cdot \widehat{u})\,(\xi \cdot \widehat{f}) + (\xi' \cdot \widehat{u})\,(\xi' \cdot \widehat{f}) = |\xi|^2 (\widehat{u} \cdot \widehat{f}), \; \xi' := (\xi_2, -\xi_1),$$

we determine

$$\|f\|'^2 = \int \langle \xi \rangle^{-1} \left(|\widehat{f}|^2 + |\xi' \cdot \widehat{f}|^2 \right) d\xi = \|h\|^2_W = \|f\|^2_{W'(\mathrm{R}^2)}.$$

Thus, norms $\|\cdot\|'$ and $\|\cdot\|_{W'(\mathrm{R}^2)}$ are equivalent. On the other hand,

$$\begin{aligned} \left| \int \widehat{f} \cdot \overline{\widehat{u}}\, d\xi \right| &= \left| \int |\xi|^{-2} \xi (\xi \cdot \overline{\widehat{u}}) \cdot \widehat{f}\, d\xi + \int |\xi|^{-2} \overline{\widehat{u}} \cdot \xi' (\xi' \cdot \widehat{f})\, d\xi \right| \\ &\le C \left(\|div\, u\|_{-1/2} \|f\|_{-1/2} + \|u\|_{-1/2} \left\| \frac{\partial f_2}{\partial x_1} - \frac{\partial f_1}{\partial x_2} \right\|_{-1/2} \right) \\ &\le C \|u\|_W \|f\|_{W'}. \end{aligned}$$

Therefore, $\int \widehat{f} \cdot \overline{\widehat{u}}\, d\xi$ is an antilinear continuous functional; consequently, spaces $W(\mathrm{R}^2)$ and $'W(\mathrm{R}^2)$ are antidual. Finally, taking into account that the density belongs to $C_0^\infty(\Omega)$ in $W(\overline{\Omega})$, we make sure that the space of restrictions of elements from $'W(\mathrm{R}^2)$ on Ω is antidual with respect to $W(\overline{\Omega})$.

The continuity of the embedding

$$H^{1/2}(\Omega) \subset W' \subset H^{-1/2}(\Omega)$$

follows from the duality relationships and Statement 8. □

Statement 11 *The space W' is expanded into a direct sum of (closed) orthogonal subspaces*

$$W' = W^1 \oplus W^2,$$

where

$$W^1 := \left\{ f \in W' : \frac{\partial f_1}{\partial x_1} + \frac{\partial f_2}{\partial x_2} = 0 \right\}$$

and

$$W^2 := \left\{ f \in W' : \frac{\partial f_2}{\partial x_1} - \frac{\partial f_1}{\partial x_2} = 0 \right\}.$$

□ Indeed, since for every $h = h_1 + h_2$, $h_1 \in W_1$ and $h_2 \in W_2$, where $h_1 \perp h_2$, then

$$(h, u)_W = (h_1, u)_{-1/2} + (h_2, u)_{1/2},$$
$$(h_1, u)_{-1/2} = \int \widehat{f}_1 \cdot \overline{\widehat{u}}\, d\xi, \quad \widehat{f}_1 = \langle \xi \rangle^{-1} \widehat{h}_1, \quad \xi \cdot \widehat{f}_1 = 0;$$
$$(h_2, u)_{1/2} = \int \widehat{f}_2 \cdot \overline{\widehat{u}}\, d\xi, \quad \widehat{f}_2 = \langle \xi \rangle \widehat{h}_2, \quad \xi' \cdot \widehat{f}_2 = 0.$$

In order to verify that W^1 and W^2 are closed and orthogonal, one can use Statements 7 and 10 and the duality relationships. □

We may consider a more general situation and introduce the scales of spaces W^s and $(W^s)'$ with the norm

$$\|u\|^2_{W^s} = \|u\|^2_s + \|div\, u\|^2_s\,, \quad s \in R.$$

However, from the physical viewpoint, such an approach gives nothing new, since our purpose is to study the (unique) continuous quasi-classical solution to the diffraction problem, and consideration of the problem in the Sobolev spaces is just a convenient method of our study. Therefore, we will consider only the space $W = W^{-1/2}$ (and, respectively, $W' = (W^{-1/2})'$).

2.3 Vector potentials and representation of solutions

Representation of fields E, H in the form of vector potentials

$$E = ik^{-1} \left(Grad\, div(A_1 u) + k^2 A_1 u \right), \tag{2.11}$$

$$H = Rot\,(A_1 u); \tag{2.12}$$

$$A_1 u = \frac{1}{4\pi} \int\limits_{\Omega} \frac{e^{ik|x-y|}}{|x-y|} u(y)\, dy; \quad x = (x_1, x_2, x_3) \in R^3 \setminus \overline{\Omega} \tag{2.13}$$

is the most natural and efficient method for solving problem (2.1)–(2.6) Here, vector $u = (u^1, u^2)^T$ is the current density on the surface of the screen Ω.

We will assume that

$$u \in W(\overline{\Omega}) \tag{2.14}$$

and

$$u,\ div\, u \in C^1(\Omega). \tag{2.15}$$

Since it is supposed that condition (2.8) holds, we will show that u is a smooth vector (and even an infinitely differentiable vector) in Ω. Therefore, from the very beginning, we can introduce a certain additional smoothness condition for u at all internal points of Ω; and (2.15) is the simplest sufficient condition which can be applied in this case. Note that condition (2.14) imposes restrictions concerning the behavior of u in the vicinity of the boundary Γ.

Integral (2.13) is a convolution of the finite generalized function u and a regular function, so that x_3 is a parameter. Let us calculate the Fourier transform (with respect to two variables) of the kernel function $|x|^{-1}\exp(ik|x|)$. Using formula (6.637) from Ref. [8], we obtain

$$\begin{aligned}\int_0^\infty \frac{e^{ik\sqrt{\rho^2+x_3^2}}}{\sqrt{\rho^2+x_3^2}} J_1(\rho|\xi|)\,d\rho &= I_{1/2}\left(\frac{|x_3|}{2}\left(\sqrt{\xi^2-k^2}+ik\right)\right) \\ &\quad \times K_{1/2}\left(\frac{|x_3|}{2}\left(\sqrt{\xi^2-k^2}-ik\right)\right) \\ &= \frac{1}{|x_3||\xi|}\left(e^{ik|x_3|}-e^{-|x_3|\sqrt{\xi^2-k^2}}\right).\end{aligned}$$

According to the properties of the Bessel function given by (801.91) from Ref. [11],

$$\rho J_0(\rho|\xi|) = \frac{d}{d|\xi|}\left(J_1(\rho|\xi|)\right) + \frac{1}{|\xi|}J_1(\rho|\xi|),$$

and formula (8.411) from Ref. [8], we determine

$$\begin{aligned}F\left(\frac{e^{ik|x|}}{|x|}\right) &= \frac{1}{2\pi}\int_0^\infty\int_0^{2\pi} \frac{e^{ik\sqrt{\rho^2+x_3^2}}}{\sqrt{\rho^2+x_3^2}} e^{-i\rho|\xi|\cos(\varphi-\psi)}\rho\,d\varphi\,d\rho \\ &= \int_0^\infty \frac{e^{ik\sqrt{\rho^2+x_3^2}}}{\sqrt{\rho^2+x_3^2}}\rho J_0(\rho|\xi|)\,d\rho \\ &= \frac{e^{-|x_3|\sqrt{\xi^2-k^2}}}{\sqrt{\xi^2-k^2}}.\end{aligned}$$

We choose the branch of the square root, for which $\Im k \geq 0$, so that

$$\zeta^{-1/2} = \frac{\sqrt{|\zeta|+\Re\zeta} + i\,sign(\Re k)\sqrt{|\zeta|-\Re\zeta}}{\sqrt{2}|\zeta|},$$
$$\zeta = \xi^2 - k^2.$$

For $\Im k > 0$, all the above transformations are justified because integrands decrease exponentially, and for real k, the formulas are extended using the continuity. From the latter expression, we obtain

$$F\left(\frac{e^{ik|x|}}{|x|}\right) = O\left(\frac{1}{|\xi|}\right) \text{ for } |\xi| \to \infty;$$

therefore,

$$\frac{e^{ik|x|}}{|x|} \in H^{1/2}(R^2)$$

and expression (2.13) is defined correctly.

If $x_3 \neq 0$, then, $A_1 u \in C^\infty(R^2)$, since (2.13) is a convolution of a finite and infinitely differentiable function. If $x_3 = 0$, then $A_1 u \in C^\infty(R^2 \setminus \overline{\Omega})$. Indeed, assume that

$x = (x_1, x_2) \notin \overline{\Omega}$ and $dist\,(x, \overline{\Omega}) > \varepsilon$. Choose $\varphi(t) \in C^\infty(\mathrm{R}^1)$ so that $0 \le \varphi \le 1$, $\varphi \equiv 1$ for $|t| < \varepsilon/2$, and $\varphi \equiv 0$ for $|t| \ge \varepsilon$. Represent (2.13) in the form

$$\begin{aligned} A_1 u &= A_1' u + A_1'' u \\ &= \frac{1}{4\pi} \int\limits_{\Omega} \varphi(|x-y|) \frac{e^{ik|x-y|}}{|x-y|} u(y)\, dy \\ &+ \frac{1}{4\pi} \int\limits_{\Omega} (1 - \varphi(|x-y|)) \frac{e^{ik|x-y|}}{|x-y|} u(y)\, dy. \end{aligned}$$

Then, $A_1'' u \in C^\infty(\mathrm{R}^2)$ as a convolution of a finite and infinitely differentiable function. In addition to this,

$$supp\, A_1' u \subset \overline{\Omega}_\varepsilon = \left\{ z \in \mathrm{R}^2 : dist\,(z, \overline{\Omega}) \le \varepsilon \right\},$$

since, according to Ref. [4], $supp\,\varphi \subset [-\varepsilon, \varepsilon]$. Hence, $A_1 u$ is infinitely differentiable at a point $x \notin \overline{\Omega}$.

The above considerations remain valid if to apply them to the derivative of $A_1 u$ with respect to x_3 of arbitrary order. Thus, we see that $A_1 u \in C^\infty(\mathrm{R}^3 \setminus \overline{\Omega})$ as a function of three variables, and the derivatives can be calculated under the integral sign in (2.13).

As has been shown in Ref. [75], operator A_1 acts continuously in the spaces

$$A_1 : \tilde{H}^{-1/2}(\overline{\Omega}) \to H^1_{loc}(\mathrm{R}^3).$$

Let us prove that

$$div\,(A_1 u) = A_1(div\, u), \quad x \notin \overline{\Omega}, \quad u \in W.$$

It is sufficient to prove this equality for functions $u \in C_0^\infty(\Omega)$, because this set is dense in W and $\tilde{H}^{-1/2}(\overline{\Omega})$. Applying the operation of divergence under the integral sign and transferring differentiation to function u, we obtain

$$\begin{aligned} div\,(A_1 u) &= \frac{1}{4\pi} \int\limits_{\Omega} u(y) \cdot grad_x \left(\frac{e^{ik|x-y|}}{|x-y|} \right) dy \\ &= -\frac{1}{4\pi} \int\limits_{\Omega} u(y) \cdot grad_y \left(\frac{e^{ik|x-y|}}{|x-y|} \right) dy \\ &= -\frac{1}{4\pi} \int\limits_{\Omega} div_y \left(\frac{e^{ik|x-y|}}{|x-y|} u(y) \right) dy + \frac{1}{4\pi} \int\limits_{\Omega} \frac{e^{ik|x-y|}}{|x-y|} div\, u\, dy \\ &= -\frac{1}{4\pi} \oint\limits_{\Gamma} \frac{e^{ik|x-y|}}{|x-y|} u \cdot n\, dl + \frac{1}{4\pi} \int\limits_{\Omega} \frac{e^{ik|x-y|}}{|x-y|} div\, u\, dy \\ &= \frac{1}{4\pi} \int\limits_{\Omega} \frac{e^{ik|x-y|}}{|x-y|} div\, u\, dy = A_1(div\, u), \quad x \notin \overline{\Omega}. \end{aligned}$$

Then, (2.11) is equivalent to

$$E = ik^{-1} \left(Grad\, A_1(div\, u) + k^2 A_1 u \right), \quad x \in \mathrm{R}^3 \setminus \overline{\Omega}. \tag{2.16}$$

Note that, for the present, condition(2.15) was not applied.

Consider the limiting transition in formulas (2.16) and (2.12) for E and H when point x is dropped on Ω. Assume that condition(2.15) is satisfied. Let $x = (x_1, x_2) \in \Omega$ be a fixed point with $dist\,(x, \Gamma) > \varepsilon$, $O_\varepsilon(x) := \{y : |y - x| < \varepsilon\}$ is the ε-vicinity of point x, $x_3 \neq 0$, and $x_3 \to 0$. We will again use the representation $A_1 u = A_1' u + A_1'' u$. The differentiation and limiting transition in the expression for $A_1'' u$ can be performed under the integral sign. For the first term, we obtain

$$\begin{aligned} A_1' u &= \frac{1}{4\pi} \int\limits_{O_\varepsilon(x)} \varphi(|x - y|) \frac{e^{ik|x-y|}}{|x - y|} u(y)\, dy \\ &= \frac{1}{4\pi} \int\limits_{O_{\varepsilon/2}(x)} \frac{e^{ik|x-y|}}{|x - y|} u(y)\, dy \\ &\quad + \frac{1}{4\pi} \int\limits_{O_\varepsilon(x) \setminus O_{\varepsilon/2}(x)} \frac{e^{ik|x-y|}}{|x - y|} \varphi(|x - y|) u(y)\, dy \\ &= \frac{1}{4\pi} I_1 + I_2. \end{aligned}$$

Since the kernel of integral I_2 is not singular, one can differentiate I_2 with respect to x_j and calculate the limit under the integral sign (here, u is a usual differentiable function). I_1 exists as a weakly singular integral, and the limit $x_3 \to 0$ can be also calculated under the integral sign.

When $x_3 \neq 0$, we obtain the following representation for derivatives $\partial/\partial x_j$:

$$\begin{aligned} \frac{\partial I_1}{\partial x_j} &= \int\limits_{\Omega_\varepsilon} \frac{e^{ik|x-y|}(ik|x - y| - 1)}{|x - y|^3} (x_j - y_j) u(y)\, dy \\ &= u(x) \int\limits_{\Omega_\varepsilon} \frac{e^{ik|x-y|}(ik|x - y| - 1)}{|x - y|^3} (x_j - y_j)\, dy \\ &\quad + \int\limits_{\Omega_\varepsilon} \frac{e^{ik|x-y|}(ik|x - y| - 1)}{|x - y|^3} (x_j - y_j)(u(y) - u(x))\, dy \end{aligned}$$

$$= u(x)\, J_1 + J_2.$$

For $j = 1, 2$, integral $J_1 = 0$, since the integrand is an odd function. For $j = 3$, we have

$$J_1 = 2\pi x_3 \left(\frac{e^{ik\sqrt{x_3^2+\varepsilon^2}}}{\sqrt{x_3^2 + \varepsilon^2}} - \frac{e^{ik|x_3|}}{|x_3|} \right)$$

and

$$\lim_{x_3 \to \pm 0} J_1 = \mp 2\pi.$$

By virtue of the Hölder inequality $|u(x) - u(y)| \leq C|x - y|$, J_2 exists as a weakly singular integral.

Thus, components E_τ and H_ν defined by formulas (2.16) and (2.12), are continuous up to Ω (except for points of the boundary Γ). The operations of differentiation and limiting transition in the corresponding expression can be performed under the integral sign, where the integral should be considered as the Cauchy singular integral. In addition to this, using the above rules, we find

$$\lim_{x_3 \to \pm 0} E_\nu = \mp \frac{i}{2k} div\, u, \quad \lim_{x_3 \to \pm 0} H_\tau = \pm \frac{1}{2} u \times e_3, \tag{2.17}$$

and, consequently,

$$[H_\tau]_\Omega = u \times e_3, \quad \nu = e_3 = (0,0,1), \tag{2.18}$$

where the symbol $[\cdot]_\Omega$ means the difference of the limiting values of a function determined for $x_3 \to +0$ and $x_3 \to -0$ at points of Ω. The direct values of the corresponding integrals vanish. The latter formula can be used for explaining the physical sense of vector u. Expression (2.17) is an analog of the corresponding relationships for single-layer and double-layer potentials.

Let us define H_τ and E_ν from each side of Ω using (2.17). Then, E and H will be continuous (from each side) in Ω and condition (2.1) will be valid.

Performing the differentiation, we see that fields $E, H \in C^\infty(\mathrm{R}^3 \setminus \overline{\Omega})$ defined by formulas (2.11)–(2.13) satisfy the Maxwell equations (2.2) in $\mathrm{R}^3 \setminus \overline{\Omega}$ (for every u).

Representing the fields in the form (2.11)–(2.13), we satisfy also conditions at infinity (2.5) and (2.6). Indeed, by virtue of (2.12) and (2.16) and the equivalence of (2.9) and (2.6), it is sufficient to check the validity of the Sommerfeld conditions

$$\frac{\partial F}{\partial r} - ikF = o(r^{-1}), \quad F = O(r^{-1}), \quad r := |x| \to \infty,$$

for the function $F = A_1 u$ and the derivatives

$$F = \frac{\partial}{\partial x_j}(A_1 u).$$

Note that, for $\Im k > 0$, one can directly verify that condition (2.5) holds. Now, we set

$$F = A_1 u, \quad \Phi = |x - y|^{-1} \exp(ik|x - y|).$$

Then,

$$\begin{aligned} \left| \frac{\partial F}{\partial r} - ikF \right| &= \left| \left(\frac{\partial \Phi}{\partial r} - ik\Phi, u \right) \right| \\ &\le \left\| \frac{\partial \Phi}{\partial r} - ik\Phi \right\|_{1/2} \|u\|_{-1/2} \\ &\le \left\| \frac{\partial \Phi}{\partial r} - ik\Phi \right\|_{1} \|u\|_{-1/2}, \end{aligned}$$

$$|F| \le \|\Phi\|_{1/2} \|u\|_{-1/2} \le \|\Phi\|_1 \|u\|_{-1/2},$$

but

$$\left\| \frac{\partial \Phi}{\partial r} - ik\Phi \right\|_{H^1(\Omega)} = o(r^{-1}), \quad \|\Phi\|_{H^1(\Omega)} = O(r^{-1}),$$

since for functions Φ and $\partial\Phi/\partial y_j$, the Sommerfeld conditions hold uniformly with respect to y in every bounded domain, in particular, in the domain $y \in \overline{\Omega}$ [45]. Hence, the Sommerfeld conditions are valid for function $A_1 u$. Similar estimates can be obtained for functions $\partial(A_1 u)/\partial x_j$.

Finally, the validity of condition (2.2), which means that the field energy is finite in every bounded spatial volume, can be proved as a direct consequence of formulas (2.12), (2.16), and (2.13) and the inclusion $A_1 u \in H^1_{loc}(\mathrm{R}^3)$. Thus, representation of fields in the form (2.11)–(2.13) ensures, subject to conditions (2.14) and (2.15), the fulfillment of all requirements that enter the formulation of diffraction problem (2.1)–(2.6) except for (2.3).

Boundary condition (2.3) yields an integrodifferential equation for u. Dropping a point x on Ω in (2.16), we obtain

$$grad\, A(div\, u) + k^2 A\, u = f, \quad x = (\mathrm{x}_1, \mathrm{x}_2) \in \Omega, \tag{2.19}$$

where

$$A\, u = \int\limits_{\Omega} \frac{e^{ik|x-y|}}{|x-y|} u(y)\, dy, \tag{2.20}$$

$$f = 4\pi i k\, E^0_\tau\Big|_\Omega, \quad f \in C^\infty(\overline{\Omega}). \tag{2.21}$$

Here, the operation *grad* is considered in R^2. We have proved the possibility of performing the limiting transition in (2.16) and (2.12) and the continuity of E_τ up to Ω; therefore, if u is a solution to (2.19) and satisfies conditions (2.14) and (2.15), then formulas (2.16) and (2.12) or (2.11)–(2.12) define the quasi-classical solution of problem (2.1)–(2.6). In addition to this, if u is a nontrivial solution, one can show, with the help of (2.17), that E, H is also a nontrivial solution to (2.1)–(2.6). Then, Theorem 1 yields the following statement.

Theorem 2 *Equation (2.19) has not more than one solution that satisfies conditions (2.14) and (2.15).*

□ If homogeneous equation (2.19) would have a nontrivial solution, formulas (2.16), (2.12) would define a nontrivial solution of problem (2.1)–(2.6), which contradicts Theorem 1. □

The possibility of representing every solution of problem (2.1)–(2.6) in the form of vector potential (2.11)–(2.13) with a certain function u is closely connected with the existence of a solution to equation (2.19) that satisfies conditions (2.14) and (2.15). Indeed, if one proves the existence of such a solution, then formulas (2.11)–(2.13) always define a unique solution of problem (2.1)–(2.6), so that every solution of (2.1)–(2.6) can be represented as a vector potential. Equation (2.19) will be investigated in Section 2.5.

We note that, in the case of a screen, one cannot use the theory of single-layer and double-layer potentials for obtaining the Fredholm equation of the second kind; such a reduction was performed in Ref. [27] for closed surfaces. Since potentials are discontinuous, the field E_τ takes different values on different sides of Ω, which yields a contradiction with boundary conditions (2.3) and the continuity of the incident field E^0_τ in (2.7),

$$[E^0_\tau]_\Omega = 0. \tag{2.22}$$

2.4 Reduction of the problem to the system of pseudodifferential equations

In this section, we investigate system of equations (2.19) in the space W : $u \in W$, $f \in C^\infty(\overline{\Omega})$. We will consider operator A, defined by formula (2.20), as a PDO [87, 94].

First, we define A on $C_0^\infty(\overline{\Omega})$. Sometimes, it will be more convenient to use another representation of operator A in the form

$$Au = \int_\Omega \frac{e^{-|x-y|}}{|x-y|} u(y)\, dy + \frac{k^2+1}{2} \int_\Omega e^{-|x-y|} u(y)\, dy \tag{2.23}$$

$$+ \int_\Omega \eta(|x-y|) \left(\frac{e^{ik|x-y|}}{|x-y|} - \frac{e^{-|x-y|}}{|x-y|} - \frac{k^2+1}{2} e^{-|x-y|} \right) u(y)\, dy,$$

where $\eta(t) = 1$ for $t \le diam\, \Omega$ and $\eta(t) = 0$ for $t \ge t_0 = 2\, diam\, \Omega$ is the infinitely differentiable cut function. Formulas (2.20) and (2.23) obviously define one and the same operator (for $x \in \Omega$, they define a restriction of A on Ω), and this definition does not depend on function η. We give two different formulas for the symbol $a(\xi)$ of PDO A,

$$Au = \int a(\xi) \widehat{u}(\xi) e^{ix\cdot\xi}\, d\xi, \quad x \in \Omega \tag{2.24}$$

(the definition of the symbol of operator A is ambiguous). Assume that $\Im k \ge 0$. (2.20) defines a convolution-type operator, and the Fourier transform of its kernel is

$$F\left(\frac{e^{ik|x|}}{|x|} \right) = \frac{1}{2\pi} \int \frac{e^{ik|x|}}{|x|} e^{-ix\cdot\xi}\, dx = \frac{1}{\sqrt{\xi^2 - k^2}};$$

therefore,

$$a(\xi) = (\xi^2 - k^2)^{-1/2}, \quad \Im k \ge 0. \tag{2.25}$$

On the other hand, by virtue of (2.23), for every $k \in \mathrm{C}$, we have

$$F\left(|x|^{-1} e^{-|x|}\right) = \langle \xi \rangle^{-1}, \quad F\left(e^{-|x|}\right) = \langle \xi \rangle^{-3},$$

$$a(\xi) = \langle \xi \rangle^{-1} + b(\xi), \quad b(\xi) = \frac{k^2+1}{2} \langle \xi \rangle^{-3} + \widehat{g}(\xi), \tag{2.26}$$

$$\widehat{g}(\xi) = Fg = F\left(\eta(|x|) \left(\frac{e^{ik|x|}}{|x|} - \frac{e^{-|x|}}{|x|} - \frac{k^2+1}{2} e^{-|x|} \right) \right). \tag{2.27}$$

For the sake of convenience, $a(\xi)$ again denotes the symbol, although it is another function. Below, we will indicate when necessary which of two representations, (2.25) or (2.26), should be applied. Usually, considerations are valid independently of the choice of the formula for the symbol. Obviously, $b \in C^\infty(\mathrm{R}^2)$. Calculating the Fourier transform (in the polar coordinates), we find

$$-i\xi_j^3 \widehat{g}(\xi) = F\left(\frac{\partial^3 g}{\partial \mathrm{x}_j^3} \right)$$

$$= \int_0^{t_0} \left(C_j^{(1)}\left(\frac{\xi}{|\xi|} \right) g_j^{(1)}(t) J_1(t|\xi|) + C_j^{(2)}\left(\frac{\xi}{|\xi|} \right) g_j^{(2)}(t) J_3(t|\xi|) \right) dt$$

with smooth functions $g_j^{(1,2)}$ and bounded $C_j^{(1,2)}$, where $J_k(t)$ are the Bessel functions and $j = 1, 2$. Functions $g_j^{(1,2)}(t)$ are absolutely integrable with the weight $t^{-1/2}$. Using asymptotic representations of the Bessel functions at infinity [8], we obtain the estimate

$$|\widehat{g}(\xi)| \le \frac{C}{\langle \xi \rangle^{7/2}}, \quad |b(\xi)| \le \frac{C}{\langle \xi \rangle^3}. \tag{2.28}$$

Since the principal part of the symbol $a(\xi)$ is $\langle\xi\rangle^{-1}$,

$$\frac{1}{\sqrt{\xi^2-k^2}} = \frac{1}{\langle\xi\rangle} + \frac{k^2+1}{2\langle\xi\rangle^3} + O\left(|\xi|^{-5}\right), \quad |\xi| \to \infty, \tag{2.29}$$

we will consider the PDO A_0 with the symbol $\langle\xi\rangle^{-1}$,

$$A_0 u = \int\limits_{\Omega} \frac{e^{-|x-y|}}{|x-y|} u(y)\,dy = \int \frac{1}{\langle\xi\rangle} \widehat{u}(\xi) e^{ix\cdot\xi}\,d\xi. \tag{2.30}$$

The quadratic form of this operator,

$$\begin{aligned}(A_0 u, u) &= \int\limits_{\Omega\times\Omega} \frac{e^{-|x-y|}}{|x-y|} \overline{u}(x) u(y)\,dx\,dy \\ &= \int \frac{1}{\langle\xi\rangle} |\widehat{u}(\xi)|^2\,d\xi = \|u\|^2_{-1/2}\end{aligned} \tag{2.31}$$

is positively defined and bounded on $\tilde{H}^{-1/2}(\overline{\Omega})$. Applying expressions (2.31) and the Lax–Milgram theorem [26, 32], we prove that

$$A_0 : \tilde{H}^{-1/2}(\overline{\Omega}) \to H^{1/2}(\Omega)$$

is an isomorphic operator on the pair of spaces $\tilde{H}^{-1/2}(\overline{\Omega})$ and $H^{1/2}(\Omega)$. The operator can be extended on entire $\tilde{H}^{-1/2}(\overline{\Omega})$ using the continuity because $C_0^\infty(\Omega)$ is dense in $\tilde{H}^{-1/2}(\overline{\Omega})$. Since $\langle\xi\rangle^2$ is the symbol of operator $1-\Delta$, where Δ is the Laplacian, we will denote by $(1-\Delta)^s$ a PDO with the symbol $\langle\xi\rangle^{2s}$ [35, 40]. Then, $A_0 = (1-\Delta)^{-1/2}$.

We define the operator L on $C_0^\infty(\Omega)$ by the formulas

$$\begin{aligned}Lu &= grad\,A\,(div\,u) + k^2 Au \\ &= \int a(\xi)\left(-\xi\,(\xi\cdot\widehat{u}(\xi)) + k^2\widehat{u}(\xi)\right) e^{ix\cdot\xi}\,d\xi\end{aligned} \tag{2.32}$$

(considering again the restriction of Lu on Ω, $x \in \Omega$). As a result, we obtain a PDO with the complete matrix symbol

$$a(\xi)\begin{pmatrix} -\xi_1^2+k^2 & -\xi_1\xi_2 \\ -\xi_1\xi_2 & -\xi_2^2+k^2 \end{pmatrix}.$$

The principal part of the matrix symbol (defined formally) is degenerate,

$$det\begin{pmatrix} -\xi_1^2 & -\xi_1\xi_2 \\ -\xi_1\xi_2 & -\xi_2^2 \end{pmatrix} = 0.$$

This circumstance does not allow us to apply the developed theory of elliptic PDOs [40, 57, 66]. In order to analyze operator L, we will use the method of splitting a PDO on subspaces. Note that the space W is coordinated with the structure of the symbol of operator L.

Let $t(u,v)$ be a bounded sesquilinear form considered on the complex space W : $|t(u,v)| \le C\|u\|_W\|v\|_W$. Then, this form uniquely defines a linear bounded operator $T : W \to W$ by the formula [26]

$$t(u,v) = (Tu, v)_W, \quad \forall v \in W. \tag{2.33}$$

It is sufficient to check that the form is bounded on $C_0^\infty(\Omega)$, since $C_0^\infty(\Omega)$ is dense in W. In addition to this, the form $t(u,v)$ itself can be completely defined on $C_0^\infty(\Omega)$.

Consider sesquilinear form (Lu, v), produced by operator L, on the space W. According to (2.32),

$$t(u,v) := -\int a(\xi)\,(\xi\cdot\hat{u}(\xi))\left(\xi\cdot\overline{\hat{v}(\xi)}\right)d\xi + k^2\int a(\xi)\hat{u}(\xi)\cdot\overline{\hat{v}(\xi)}\,d\xi, \tag{2.34}$$

with the symbol $a(\xi)$ specified by formulas (2.25) and (2.26).

Statement 12 *Form (2.34) defines a linear bounded operator $T : W \to W$ by formula (2.33).*

□ The proof follows from the estimate

$$\begin{aligned} |t(u,v)| &\le C\int \langle\xi\rangle^{-1}\,|\hat{u}(\xi)|\,|\hat{v}(\xi)|\,d\xi + C\int \langle\xi\rangle^{-1}\,|\xi\cdot\hat{u}(\xi)|\,|\xi\cdot\hat{v}(\xi)|\,d\xi \\ &\le C\|u\|_W\|v\|_W. \end{aligned}$$

□

Continuity of form (2.34) in W enables us to consider L as a bounded operator $L : W \to W'$, where W' is the antidual space of W:

$$t(u,v) = (Lu, v), \ \ \forall v \in W, \quad Lu \in W', \ \ u \in W, \tag{2.35}$$

$$\|Lu\|_{W'} = \sup_{0\neq v\in W}\frac{|t(u,v)|}{\|v\|_W} \le C\|u\|_W.$$

For functions $u \in C_0^\infty(\Omega)$, operator L is defined in a usual manner. Since $C_0^\infty(\Omega)$ is dense in W, operator L can be extended on W using the continuity.

Now we can consider (2.19) as a system of pseudodifferential equations

$$Lu = f, \quad u \in W, \quad f \in C^\infty(\overline{\Omega}) \subset W'. \tag{2.36}$$

The equality in (2.36) is understood in the sense of distributions: multiplying (2.36) by an arbitrary element $\overline{v} \in C_0^\infty(\Omega)$ and integrating over Ω, we obtain the variational identity

$$\begin{aligned} -\int a(\xi)\,(\xi\cdot\hat{u}(\xi))\left(\xi\cdot\overline{\hat{v}(\xi)}\right)d\xi + k^2\int a(\xi)\hat{u}(\xi)\cdot\overline{\hat{v}(\xi)}\,d\xi \\ = \int_\Omega f(x)\cdot\overline{v(\xi)}\,dx. \end{aligned} \tag{2.37}$$

Definition 2 *An element $u \in W$ is called the generalized solution to system of equations (2.36) or (2.19), if, for every $v \in C_0^\infty(\Omega)$, variational identity (2.37) is valid.*

Thus, we consider equation (2.19) on a space, which is wider than that specified by conditions (2.14) and (2.15), so that condition (2.15) is not applied. It is much easier to analyze equation (2.19) on W. In Section 2.5, we will prove the solvability of this equation in the space W. Then, in Section 2.6, we will show that any generalized solution $f \in C^\infty(\overline{\Omega})$ satisfies condition (2.15).

2.5 Fredholm property and solvability of the system of pseudodifferential equations

The proof that T and L are Fredholm operators with a zero index (in the corresponding spaces) constitutes a central part of this section. We recall that a bounded operator F with a closed domain of values is called the Fredholm operator, if $dim\, ker\, F < \infty$ and $dim\, coker\, F < \infty$; the difference $ind\, F = dim\, ker\, F - dim\, coker\, F$ is the index of operator F. In order to prove that F is a Fredholm operator with a zero index, it is sufficient to show that $F = S + K$, where S is a continuously invertible operator and K is a compact operator [25, 26]). We will prove that T and L are Fredholm operators using an important property of form (2.34), which can be directly verified: $t(u,v) = 0$ when $u \in W_i$ and $v \in W_j$ for $i \neq j$,

Statement 13 *Operator $T = T(k) : W \to W$ defined by form (2.34) is a Fredholm operator for $k \neq 0$, and $ind T = 0$.*

□ Let $u \in W_1$ and $v \in W_2$. Then, according to Statement 7, we have

$$(Tu, v)_W = t(u,v) = 0, \quad Tu \perp W_2, \quad Tu \in W_1$$

Thus, W_1 is an invariant subspace for operator T. Performing transposition of indices $1 \leftrightarrow 2$, we prove that W_2 is also an invariant space for T. As a result, we obtain the expansion

$$T = \begin{pmatrix} T_1 & 0 \\ 0 & T_2 \end{pmatrix} : \begin{pmatrix} W_1 \\ W_2 \end{pmatrix} \to \begin{pmatrix} W_1 \\ W_2 \end{pmatrix}$$

with operators $T_j : W_j \to W_j$, $j = 1, 2$, produced by the restriction of form $t(u,v)$ on W_j:

$$(T_1 u, v)_{W_1} = t_1(u,v) = k^2 \int a(\xi)\hat{u}(\xi) \cdot \overline{\hat{v}(\xi)}\, d\xi, \quad \forall v \in W_1,$$

$$(T_2 u, v)_{W_2} = t_2(u,v) = \int a(\xi)(k^2 - \xi^2)\hat{u}(\xi) \cdot \overline{\hat{v}(\xi)}\, d\xi, \quad \forall v \in W_2.$$

Let us prove the Fredholm property of each operator T_j.

Consider first T_1. It is convenient to represent t_1 in the form

$$t_1(u,v) = k^2 \int \frac{1}{\langle\xi\rangle}\hat{u}(\xi) \cdot \overline{\hat{v}(\xi)}\, d\xi + \int b_1(\xi)\hat{u}(\xi) \cdot \overline{\hat{v}(\xi)}\, d\xi,$$

where $b_1 \in C^\infty(\mathrm{R}^2)$ and, by virtue of (2.28), $|b_1(\xi)| \leq C\langle\xi\rangle^{-3}$. The first and the second terms of the form define, respectively, operator $k^2 I$, where I is the unit operator, and a compact operator $B_1 : W_1 \to W_1$. Indeed, the following estimate is valid

$$\left|(B_1 u, v)_{W_1}\right| \leq C \|u\|_{-5/2} \|v\|_{-1/2} = C \|u\|_{-5/2} \|v\|_{W_1} \cdot$$

Setting $v = B_1 u$, we obtain $\|B_1 u\|_{W_1} \leq C \|u\|_{-5/2}$. Let $u_n \to u$ weakly in W_1; then, using the continuity of the embedding

$$W_1 \subset \tilde{H}^{-1/2}(\overline{\Omega})$$

(Statement 8) and compactness of the embedding [58]

$$\tilde{H}^{-1/2}(\overline{\Omega}) \subset \tilde{H}^{-5/2}(\overline{\Omega}),$$

we prove that $u_n \to u$ strongly in $\tilde{H}^{-5/2}(\overline{\Omega})$. From the latter estimate, it follows that $B_1 u_n \to B_1 u$ strongly in W_1. The compactness of B_1 is proved.

Consider operator T_2. Represent t_2 in the form

$$t_2(u,v) = -\int \langle\xi\rangle \hat{u}(\xi)\cdot\overline{\hat{v}(\xi)}\,d\xi + \int b_2(\xi)\hat{u}(\xi)\cdot\overline{\hat{v}(\xi)}\,d\xi,$$

where $b_2 \in C^\infty(\mathrm{R}^2)$ and, by virtue of (2.28), $|b_2(\xi)| \le C\langle\xi\rangle^{-1}$. The first term defines operator $-I$, and the second term, a compact operator $B_2 : W_2 \to W_2$. To prove the compactness, it is sufficient to verify the inequality

$$\left|(B_2 u, v)_{W_2}\right| \le C\,\|u\|_{-3/2}\,\|v\|_{W_2}$$

and to repeat the above considerations, taking into account the continuity of the embeddings

$$W_2 \subset \tilde{H}^{1/2}(\overline{\Omega}) \subset \tilde{H}^{-3/2}(\overline{\Omega}),$$

where the last embedding is compact [58]. Here, we also apply the relationships for norms that enter Statement 8.

Thus, we have obtained the expansion

$$T = \begin{pmatrix} k^2 I + B_1 & 0 \\ 0 & -I + B_2 \end{pmatrix} : \begin{pmatrix} W_1 \\ W_2 \end{pmatrix} \to \begin{pmatrix} W_1 \\ W_2 \end{pmatrix} \tag{2.38}$$

with compact operators B_1 and B_2. We will also use matrix designations

$$\begin{aligned} T &= \alpha I + B : W \to W, \\ \alpha &= diag(k^2, -1), \\ \mathcal{B} &= diag(B_1, B_2). \end{aligned} \tag{2.39}$$

Obviously, αI is a continuously invertible operator and B is a compact operator. Therefore, T is a Fredholm operator and $ind\,T = 0$. □

Remark 1 *For $k = 0$, $T(0)$ is not a Fredholm operator, since $W_1 \subset ker\,T(0)$ and $dim\,ker\,T(0) = \infty$.*

Statement 14 *Operator $T = T(k) : W \to W$ is continuously invertible for $\Im k \ge 0$ and $k \ne 0$.*

□ By virtue of Statement 13 and the Fredholm alternative, it is sufficient to show that $ker\,T = \{0\}$, or, in its turn, that $\Im(k^{-1}t(u,u)) > 0$ for $u \ne 0$. Let $u = u_1 + u_2 \ne 0$, $u_1 \in W_1$, and $u_2 \in W_2$. Using representation (2.25) for symbol $a(\xi)$ and setting $\zeta = \xi^2 - k^2$, we obtain

$$\begin{aligned} I &= \Im(k^{-1}t(u,u)) = \Im(k^{-1}t(u_1,u_1)) + \Im(k^{-1}t(u_2,u_2)) \\ &= \Im\left(k\int \frac{1}{\sqrt{\zeta}}|\hat{u}_1(\xi)|^2\,d\xi\right) + \Im\left(-\frac{1}{k}\int \sqrt{\zeta}\,|\hat{u}_2(\xi)|^2\,d\xi\right) \\ &= \int\left(\Im\frac{k}{\sqrt{\zeta}}\right)|\hat{u}_1(\xi)|^2\,d\xi + \int\left(\Im\frac{k}{\sqrt{\zeta}}\right)\frac{|\zeta|}{|k|^2}\,|\hat{u}_2(\xi)|^2\,d\xi. \end{aligned}$$

For $|k| \neq |\xi|$, we have

$$\Im \frac{k}{\sqrt{\zeta}} = \frac{|\Re k|\sqrt{|\zeta| - \Re\zeta} + |\Im k|\sqrt{|\zeta| + \Re\zeta}}{\sqrt{2}|\zeta|}.$$

Obviously, $I > 0$ for $\Im k > 0$. For $\Im k = 0$ and $k \neq 0$, we obtain

$$\begin{aligned} I &= |k| \int\limits_{|\xi|<|k|} \frac{1}{\sqrt{|k|^2 - \xi^2}} |\widehat{u}_1(\xi)|^2 \, d\xi \\ &+ \frac{1}{|k|} \int\limits_{|\xi|<|k|} \sqrt{|k|^2 - \xi^2} \, |\widehat{u}_2(\xi)|^2 \, d\xi \geq 0. \end{aligned}$$

If $I = 0$, then $\widehat{u}(\xi) = 0$ for $|\xi| < |k|$. Since $\widehat{u}(\xi)$ is the Fourier transform of a distribution with a compact support, we have $\widehat{u}(\xi) \in C^\infty(\mathrm{R}^2)$, and $\widehat{u}(\xi)$ can be the continued as an entire function $\widehat{u}(z)$ on C^2, where $z = \xi + i\eta$, and $\widehat{u}(z)|_{\eta=0} = \widehat{u}(\xi)$. Then, the condition $\widehat{u}(\xi) = 0$ in the circle $|\xi| < |k|$ yields $\widehat{u}(z) = 0$ [64] and, consequently, $u = 0$ (as an element of W), which contradicts the condition $u \neq 0$. Hence, $I > 0$ also in the case $\Im k = 0$ and $k \neq 0$. The proof of the statement is completed. □

Using Statements 8 and 9, we can prove the following theorem.

Theorem 3 *For $k \neq 0$, operator (2.32) $L = L(k) : W \to W'$ is a Fredholm operator continuously invertible for $\Im k \geq 0$ and $k \neq 0$ and $\operatorname{ind} L = 0$.*

For a function $u \in C_0^\infty(\Omega)$, $u = u_1 + u_2$, where $u_1 \in W_1$ and $u_2 \in W_2$, one can directly verify (applying, as well as in Section 2.2, the formula of integration by parts) that $L u_1 \in W^1$ and $L u_2 \in W^2$. Then, by virtue of the continuity of operator L, we obtain, taking into account that subspaces W^1 and W^2 are closed (Statement 11) that $L : W_1 \to W^1$ and $L : W_2 \to W^2$, which yields the expansion

$$L = \begin{pmatrix} L_1 & 0 \\ 0 & L_2 \end{pmatrix} : \begin{pmatrix} W_1 \\ W_2 \end{pmatrix} \to \begin{pmatrix} W^1 \\ W^2 \end{pmatrix}. \tag{2.40}$$

Operators T and L are coupled by the relationship $L = JT$, where $J : W \to W'$ is the isomorphism:

$$J = \begin{pmatrix} (1-\Delta)^{-1/2} & 0 \\ 0 & (1-\Delta)^{1/2} \end{pmatrix} : \begin{pmatrix} W_1 \\ W_2 \end{pmatrix} \to \begin{pmatrix} W^1 \\ W^2 \end{pmatrix}, \tag{2.41}$$

$$\begin{aligned} L_1 &= (1-\Delta)^{-1/2} T_1 : W_1 \to W^1, \\ L_2 &= (1-\Delta)^{1/2} T_2 : W_2 \to W^2. \end{aligned} \tag{2.42}$$

Now, the assertions of the theorem follow from (2.40)–(2.42) and Statements 8 and 9 applied for operator T. □

We can prove an important corollary from Theorem 3.

Corollary 1 *For $\Im k \geq 0$ and $k \neq 0$, there exists a unique (generalized) solution $u \in W$ of equation (2.19) or (2.36) for arbitrary right-hand side $f \in W'$ and, in particular, for $f \in C^\infty(\overline{\Omega})$.*

Remark 2 *Note that, unlike the approach used in Chapter 3 and in Refs. [94, 95], we have proved Theorem 2 without applying the uniqueness theorem for diffraction problem (2.1)–(2.6). In fact, we succeeded in establishing that* $\ker T = \{0\}$ *for* $\Im k \geq 0$ *and* $k \neq 0$ *only on the basis of the analysis of the quadratic form* $t(u,u)$.

Now, let us prove an additional statement, which will be used in Section 2.6.

Lemma 4 *Operators*

$$K_1 := (1-\Delta)^{-1/2} B_1 : W_1 \to H^{5/2}(\Omega)$$

and

$$K_2 := (1-\Delta)^{1/2} B_2 : W_2 \to H^{3/2}(\Omega)$$

are bounded.

□ For $u \in W_1$,

$$\begin{aligned} \|K_1 u\|_{H^{5/2}(\Omega)} &= \sup_{v \in \tilde{H}^{-5/2}(\overline{\Omega})} \frac{|(K_1 u, v)|}{\|v\|_{-5/2}} \\ &= \sup_{v \in \tilde{H}^{-5/2}(\overline{\Omega})} \frac{|(B_1 u, v)_{W_1}|}{\|v\|_{-5/2}} \leq C\|u\|_{W_1}. \end{aligned}$$

For $u \in W_2$,

$$\begin{aligned} \|K_2 u\|_{H^{3/2}(\Omega)} &= \sup_{v \in \tilde{H}^{-3/2}(\overline{\Omega})} \frac{|(K_2 u, v)|}{\|v\|_{-3/2}} \\ &= \sup_{v \in \tilde{H}^{-3/2}(\overline{\Omega})} \frac{|(B_2 u, v)_{W_2}|}{\|v\|_{-3/2}} \leq C\|u\|_{W_2}. \end{aligned}$$

The latter inequalities are similar to the estimates of Statement 8. □

2.6 Smoothness of generalized solutions. The orders of singularity in the vicinities of edges

In the previous section, we have proved the unique solvability of the generalized solution of equation (2.19) for every $f \in W'$. However, in (2.7)–(2.8), we have assumed that $f \in C^\infty(\overline{\Omega}) \subset W'$ (since all field sources are situated outside the screen $\overline{\Omega}$). It is natural to expect that, in this case, the solution u of equation (2.19) will be also smooth.

In order to make sure that the latter property of the solution is valid, let us take $f \in C^\infty(\overline{\Omega})$ and represent f in the form (see Statement 11):

$$\begin{gathered} f = f^1 + f^2, \qquad f^1, f^2 \in C^\infty(\overline{\Omega}), \\ \frac{\partial f_1^1}{\partial x_1} + \frac{\partial f_2^1}{\partial x_2} = 0, \qquad \frac{\partial f_1^2}{\partial x_2} - \frac{\partial f_2^2}{\partial x_1} = 0. \end{gathered}$$

The form $t(u,v)$ that produces operator L, satisfies the following conditions: $t(u,v) = 0$ for $u \in W_i$, $v \in W_j$, and $i \neq j$; therefore, system (2.19) splits into two independent equations similar to (2.40)

$$\begin{aligned} k^2 \int a(\xi)\hat{u}_1(\xi) e^{ix\cdot\xi}\, d\xi &= f^1, \quad u_1 \in W_1, \\ \int a(\xi)(k^2 - \xi^2)\hat{u}_2(\xi) e^{ix\cdot\xi}\, d\xi &= f^2, \quad u_2 \in W_2, \end{aligned}$$

or, in other notations, according to (2.41)–(2.42),

$$L_1 u_1 \equiv k^2(1-\Delta)^{-1/2} u_1 + K_1 u_1 = f^1, \tag{2.43}$$

$$L_2 u_2 \equiv -(1-\Delta)^{1/2} u_2 + K_2 u_2 = f^2. \tag{2.44}$$

Lemma 4 yields

$$K_1 u_1 \in H^{5/2}(\Omega), \quad K_2 u_1 \in H^{3/2}(\Omega),$$

and one can rewrite (2.43) and (2.44) in the form (for $(k \neq 0)$

$$(1-\Delta)^{-1/2} u_1 = g_*^1, \quad g_*^1 \in H^{5/2}(\Omega), \tag{2.45}$$

$$(1-\Delta)^{1/2} u_2 = g_*^2, \quad g_*^2 \in H^{3/2}(\Omega). \tag{2.46}$$

Operators L_1 and L_2 are PDOs of the orders -1 and 1 and, according to Refs. [15, 40], they belong to spaces $L^{-1}(\Omega)$ and $L^1(\Omega)$, respectively. Indeed, for their symbols,

$$a_1(\xi) = k^2 a(\xi), \quad a_2(\xi) = (k^2 - \xi^2) a(\xi),$$

where $a(\xi)$ is given by formulas (2.26) and (2.27), the estimates the estimate

$$\left| D_\xi^\alpha a_1(\xi) \right| \leq C_\alpha \langle \xi \rangle^{-1-|\alpha|}, \tag{2.47}$$

$$\left| D_\xi^\alpha a_2(\xi) \right| \leq C_\alpha \langle \xi \rangle^{1-|\alpha|}, \tag{2.48}$$

are valid for every multiindex

$$\alpha = (\alpha_1, \alpha_2); \quad D_\xi^\alpha = \left(\frac{1}{i} \frac{\partial}{\partial \xi_1} \right)^{\alpha_1} \left(\frac{1}{i} \frac{\partial}{\partial \xi_2} \right)^{\alpha_2}, \quad |\alpha| = \alpha_1 + \alpha_2.$$

Let us verify (2.47) and (2.48). For the function $\widehat{g}(\xi)$ defined in (2.27), we find

$$D_\xi^\alpha \widehat{g}(\xi) \frac{1}{2\pi} D_\xi^\alpha \int\limits_{|x| \leq t_0} g(|x|) e^{-ix\cdot\xi}\, dx$$

$$= \frac{1}{2\pi} \int\limits_{|x| \leq t_0} g(|x|)(-x)^\alpha e^{-ix\cdot\xi}\, dx$$

$$= \frac{1}{2\pi} \int\limits_0^{2\pi} d\varphi \int\limits_0^{t_0} g(t)(-1)^{|\alpha|} \cos^{\alpha_1}\varphi \sin^{\alpha_2}\varphi\, t^{1+|\alpha|} e^{-it|\xi|\cos(\varphi-\psi)}\, dt$$

$$= \sum_{0 \leq n \leq |\alpha|} C_n(\psi) \int\limits_0^{t_0} g(t)\, t^{1+|\alpha|} J_n(t|\xi|)\, dt,$$

where $C_n(\psi)$ are trigonometric polynomials and ψ is the polar angle of point ξ. Taking into account that $g(t) \in C^\infty[0,\infty)$ is a finite function with a support $[0, t_0]$ and using the formula

$$\frac{d}{dt}\left(t^{n+1} J_{n+1}(t|\xi|) \right) = |\xi|\, t^{n+1} J_n(t|\xi|), \quad n \geq 0,$$

we perform integration by parts $|\alpha| + 1$ times in the last term to obtain

$$\int\limits_0^{t_0} g(t)\, t^{1+|\alpha|} J_n(t|\xi|)\, dt = O\left(|\xi|^{-3/2-|\alpha|} \right), \quad |\xi| \to \infty,$$

and, consequently,

$$\left|D_\xi^\alpha \widehat{g}(\xi)\right| \le C_\alpha \langle\xi\rangle^{-3/2-|\alpha|}.$$

For $\langle\xi\rangle^s$, $s \in \mathrm{R}$, we have the obvious estimates

$$\left|D_\xi^\alpha \langle\xi\rangle^s\right| \le C_\alpha \langle\xi\rangle^{s-|\alpha|}.$$

Then, from (2.26) and (2.27), we obtain inequalities (2.47) and (2.48).

Operators L_1 and L_2 are elliptic PDOs [15, 40]; that is, in line with estimates (2.28),

$$|a_1(\xi)| \ge C_1 \langle\xi\rangle^{-1}, \quad |a_2(\xi)| \ge C_2 \langle\xi\rangle$$

for $|\xi| \ge R$ for a certain R. Existence of the parametrix for an elliptic PDO belonging to the class $L^m(\Omega)$ yields regularity of solutions to equations (2.43) and (2.44) with smooth right-hand sides [15, 40]: $u_1, u_2 \in C^\infty(\Omega)$ because f^1, $f^2 \in C^\infty(\overline{\Omega})$. Thus, we have proved

Statement 15 *If $u \in W$ is a solution to equation (2.19) with a smooth right-hand side $f \in C^\infty(\overline{\Omega})$, then $u \in C^\infty(\Omega)$.*

Note that here, the smoothness of a solution is proved only inside the domain Ω, and nothing is asserted about the smoothness up to the boundary. The study of the smoothness of generalized solutions in the vicinity of the boundary (including boundary points) is a much more complicated problem, especially in the vicinity of corner points of the boundary (edges). For differential operators, these problems were investigated in the classical monograph [28]. For the PDO $\Delta^{\pm 1/2}$, exact estimates were obtained in Ref. [88]. Below, we will apply the results of Ref. [88] and the possibility of reducing the problem under consideration to equations (2.45) and (2.46) to estimate the order of singularity of solutions in the vicinity of corner points. We will separate two cases: behavior of a solution in the vicinity of a smooth part of the boundary Γ and in the vicinity of a corner point.

Let us formulate several additional statements.

Lemma 5 *Let $f \in H^{s\pm 1}(\Omega)$ and $u \in \tilde{H}^{\mp 1/2}(\overline{\Omega})$ be a solution to the equation*

$$(1-\Delta)^{\mp 1/2} u = f, \quad 0 < s \pm 1/2 < 1/2.$$

Then, $u \in \tilde{H}^s(\overline{\Omega})$.

□ This lemma is a corollary of Theorem 2.5 proved in Ref. [88]. Obviously, f and u may be both scalar and vector functions. □

Denote by $K(p,0)$ a linear operator defined in Ref. [88] by the formula

$$(K(p,0)v)(\mathrm{x}_1,\mathrm{x}_2) = \int e^{i\mathrm{x}_1\xi_1} \langle\xi_1\rangle^{1/2} \omega(\langle\xi_1\rangle \mathrm{x}_2)(\langle\xi_1\rangle \mathrm{x}_2)^p \widehat{v}(\xi_1)\, d\xi_1, \tag{2.49}$$

where $v \in H^s(\mathrm{R}^1)$, $\langle\xi_1\rangle = (1+\xi_1^2)^{1/2}$, $(\mathrm{x}_1,\mathrm{x}_2) \in \mathrm{R}_+^2$, and $\omega(t)$ is the infinitely differentiable cut function equal unity for $0 \le t \le 1$ and zero for $t > 2$. In the operator designation applied in Ref. [88], $K(p,0)$ means that logarithmic-singular terms are absent. For the elements $v \in H^s(\Gamma)$, $K(p,0)$ can be defined in a similar manner as a linear operator, $(K(p,0)v)(x) \in \Omega$, with the help of the partition of unity and under the assumption that the boundary Γ of the domain Ω is smooth.

Lemma 6 *Assume that* Γ *is a smooth curve,* $f \in H^{s\pm 1}(\Omega)$*, and* $u \in \tilde{H}^{\mp 1/2}(\overline{\Omega})$ *is a solution to the equation* $(1-\Delta)^{\mp 1/2}u = f$*. If* $1/2 < s \pm 1/2 < 1$*, then*

$$u = u_0 + K(\mp 1/2, 0)v,$$

with $u_0 \in \tilde{H}^s(\overline{\Omega})$ *and* $v \in H^s(\Gamma)$*. If* $1 < s \pm 1/2 < 3/2$*, then*

$$u = u_0 + K(-1/2, 0)v$$

for a solution to the equation $(1-\Delta)^{-1/2}u = f$ *and*

$$u = u_0 + K(1/2, 0)v + K(1, 0)w$$

for a solution to the equation $(1-\Delta)^{1/2}u = f$*, where* $u_0 \in \tilde{H}^s(\overline{\Omega})$ *and* $v, w \in H^s(\Gamma)$*.*

□ The statements of the lemma are simple corollaries of Theorems 3.5 [87] and 5.2 [88].
□

The terms of $K(p,0)v$ define the order of singularity in the vicinity of the boundary. If $v \in H^s(\mathrm{R}^1)$, then, by virtue of (2.49), we have

$$(K(p,0)v)(\mathrm{x}_1, \mathrm{x}_2) \sim \mathrm{x}_2^p\, g(\mathrm{x}_1) \quad \text{for } \mathrm{x}_2 \to +0$$

with a function $g \in H^{s+p+1/2}(\mathrm{R}^1)$. If $v \in H^s(\Gamma)$, then

$$(K(p,0)v)(x) \sim \rho^p(x)\, g(t), \quad \rho(x) = dist(x, \Gamma) = |x - t| \tag{2.50}$$

in the vicinity of Γ, and $g \in H^{s+p+1/2}(\Gamma)$, $t \in \Gamma$.

Let us proceed to the analysis of the smoothness of solutions to equations (2.45) and (2.46). Since $g_*^1 \in H^{5/2}(\Omega)$ and $g_*^2 \in H^{3/2}(\Omega)$, the smoothness conditions for the right-hand side in Lemmas 5 and 6 are fulfilled. Lemma 5 provides the best result in nonweighted Sobolev spaces:

$$u_1 \in \tilde{H}^{-\varepsilon}(\overline{\Omega}), \quad u_2 \in \tilde{H}^{1-\varepsilon}(\overline{\Omega}),$$

where $\varepsilon > 0$ is an arbitrarily small number. Thus, there exists the trace

$$u_2|_\Gamma \in H^{1/2-\varepsilon}(\Gamma)$$

(according to the trace theorem [32]), and

$$u_2|_\Gamma = 0,$$

since the zero continuation of u_2 outside Ω is continuous with respect to $\|\cdot\|_{1-\varepsilon}$ norm. The index in the Sobolev spaces increases by $1/2 - \varepsilon$ as compared with Statement 9.

If the boundary Γ is smooth, then, by virtue of Lemma 6, we obtain

$$u_1 = u_1^0 + K(-1/2, 0)v^1,$$

$$u_1^0 \in \tilde{H}^{1-\varepsilon}(\overline{\Omega}), \quad v^1 \in H^{1-\varepsilon}(\Gamma), \tag{2.51}$$

$$u_2 = u_2^0 + K(1/2, 0)v^2 + K(1, 0)w^2,$$

$$u_2^0 \in \tilde{H}^{2-\varepsilon}(\overline{\Omega}), \quad v^2, w^2 \in H^{2-\varepsilon}(\Gamma). \tag{2.52}$$

From (2.51), it follows that, in the vicinity of the boundary, function u_1 has a singularity of the form

$$u_1 \sim \rho^{-1/2}(x)g^1(t), \quad g^1 \in H^{1-\varepsilon}(\Gamma), \tag{2.53}$$

since, by the trace theorem,

$$u_1^0\Big|_\Gamma \in H^{1/2-\varepsilon}(\Gamma) \text{ and } u_1^0\Big|_\Gamma = 0.$$

Using (2.52) we find that, in the vicinity of Γ, u_2 has a singularity

$$u_2 \sim \rho^{1/2}(x)g^2(t) + \rho(x)q^2(t), \quad g^2 \in H^{3-\varepsilon}(\Gamma),\ q^2 \in H^{7/2-\varepsilon}(\Gamma). \tag{2.54}$$

According to the embedding theorem [57], function u_2^0 satisfies in $\overline{\Omega}$ the Hölder inequality with the index $1-\varepsilon$; therefore, in the vicinity of Γ, this function has a singularity of an order of $O(\rho^{1-\varepsilon}(x))$ $(1-\varepsilon > 1/2)$.

If the boundary has corner points, then the behavior of u_1 and u_2 in the vicinity of every smooth part Γ' of the boundary Γ (situated at a certain distance from the corner points) is also described by (2.53) and (2.54), where Γ should be replaced by Γ'. In order to prove this fact, one has to cut u in the vicinity of Γ' (see Ref. [88]).

Now we assume that $n = n(t)$ is the vector of the external normal to the boundary at a point $t \in \Gamma'$. Applying (2.51)-(2.54) we find that, in the vicinity of a smooth part Γ',

$$u_2 \cdot n|_{\Gamma'} = 0, \quad u_1^0 \cdot n\Big|_{\Gamma'} = 0. \tag{2.55}$$

In addition to this, $div\, u_1 = 0$; therefore,

$$div\,(K(-1/2,0)v^1) = -div\, u_1^0 \in \tilde{H}^{-\varepsilon}(\overline{\Omega}). \tag{2.56}$$

Let Γ' be a rectilinear part of the boundary. Without loss of generality, one can assume that Γ' lies on the axis $O\mathrm{x}_1$ and the direction of the normal vector on Γ' is opposite to the direction of $O\mathrm{x}_2$. Let

$$\chi(x) = \chi_1(\mathrm{x}_1)\chi_2(\mathrm{x}_2) \in C_0^\infty(\mathrm{R}^2) \quad (0 \le \chi \le 1)$$

be the cut function equal to zero outside a certain vicinity of Γ', so that $\chi_1(\mathrm{x}_1) = 1$ for $\mathrm{x}_1 \in \Gamma'$ and $\chi_2(\mathrm{x}_2) = 1$ for $|\mathrm{x}_2| < \delta$ for a certain δ. Consider the Fourier transform of the element $h = (h_1, h_2)^T$,

$$h(x) = \chi_2(\mathrm{x}_2)\theta^{-1/2}(\mathrm{x}_2)\chi_1(\mathrm{x}_1)g^1(\mathrm{x}_1),$$

where $g^1 = (g_1^1, g_2^1)^T \in H^{1-\varepsilon}(\mathrm{R}^1)$ is a vector, and

$$\theta(\mathrm{x}_2) = \begin{cases} \mathrm{x}_2, & \mathrm{x}_2 \ge 0, \\ 0, & \mathrm{x}_2 < 0. \end{cases}$$

We have

$$\chi_1 g^1 \in H^{1-\varepsilon}(\mathrm{R}^1), \quad \widehat{\chi_2\theta^{-1/2}}(\xi_2) \sim C|\xi_2|^{-1/2}$$

for $|\xi_2| \to \infty$; therefore, $\chi_2\theta^{-1/2} \in H^{-\varepsilon}(\mathrm{R}^1)$. Then,

$$\frac{\partial h_1}{\partial \mathrm{x}_1} \in \tilde{H}^{-\varepsilon}(\overline{\mathrm{R}_+^2}).$$

On the other hand,

$$\frac{\partial h_2}{\partial x_2} \notin \tilde{H}^{-\varepsilon}(\overline{\mathbb{R}^2_+}),$$

because

$$\left\|\frac{\partial h_2}{\partial x_2}\right\|_s^2 = \iint \left|\widehat{\chi_1 g_2^1}(\xi_1)\right|^2 |\xi_2|^2 \left|\chi_2 \widehat{\theta^{-1/2}}(\xi_2)\right|^2 (1+\xi_1^2+\xi_2^2)^s \, d\xi_1\, d\xi_2 < \infty$$

if and only if $s < -3/4$. Hence, (2.56) holds only when $g_2^1(x_1) = 0$, $x_1 \in \Gamma'$. Combining this result with (2.55) we see that the normal component of u on the boundary Γ' vanishes.

If Γ' is a smooth but not a rectilinear part of the boundary, we introduce in the vicinity of Γ' curvilinear orthogonal coordinates: ρ is the distance from point x to Γ' (with the minus sign if the point lies on the side of the normal to Γ') and t is the natural parameter on Γ', counted from a certain fixed point on Γ' to the projection of x on Γ'. The first quadratic form is given by the expression

$$ds^2 = d\rho^2 + (1+\kappa(t)\rho)^2\, dt^2,$$

where $\kappa(t)$ is the curvature of Γ'. In the local coordinates,

$$div\, h = \frac{\partial h_\rho}{\partial \rho} + \frac{1}{1+\kappa(t)\rho}\left(\frac{\partial h_t}{\partial t} + \kappa(t) h_\rho\right). \tag{2.57}$$

As can be seen from (2.57), for sufficiently small ρ ($|\rho| < \delta$ for a certain positive δ), we can repeat the above considerations for the vector $h = (h_\rho, h_t)^T$,

$$h(x) = \chi_2(\rho)\theta^{-1/2}(\rho)\chi_1(t) g^1(t), \quad g^1 = (g_\rho^1, g_t^1)^T \in H^{1-\varepsilon}(\Gamma'),$$

to obtain

$$\frac{1}{1+\kappa(t)\rho}\left(\frac{\partial h_t}{\partial t} + \kappa(t) h_\rho\right) \in \tilde{H}^{-\varepsilon}(\overline{\Omega}),$$

while

$$\frac{\partial h_\rho}{\partial \rho} \notin \tilde{H}^{-\varepsilon}(\overline{\Omega}).$$

We see that $g_\rho^1(t) = 0$ on Γ' and, consequently, $h_\rho = 0$. Taking into account (2.55), we find that the normal component of u on Γ' equals zero. Thus, we have proved

Statement 16 *If Γ is a smooth curve, then $u \cdot n|_\Gamma = 0$ as an element of the space $H^{1/2-\varepsilon}(\Gamma)$. If Γ' is a smooth part of Γ situated at a positive distance from the corner points, then $u \cdot n|_{\Gamma'} = 0$ as an element of the space $H^{1/2-\varepsilon}(\Gamma')$.*

Let us analyze the singularity of u in the vicinity of the corner point P. Let $\alpha = \alpha(P)$ be the internal (with respect to Ω) angle between two smooth arcs of the boundary that intersect at the point P, such that $0 < \alpha(P) < 2\pi$ and $\alpha(P) \neq \pi$. Since $u_2|_\Gamma = 0$, only the function u_1 may be singular. Let us perform a diffeomorphic mapping of the vicinity of P on the δ-vicinity of the origin so that the arcs that form the angle would turn into the rectilinear beams $\varphi = 0$ and $\varphi = \alpha$ in the local polar coordinates (r, φ) of this δ-vicinity. Since u_1 is a solution of equation (2.46), it follows from Ref. [88] that, in the vicinity of point P, the singularity has the form

$$u_1 \sim r^{-\tau}\varphi^{-1/2}(\alpha-\varphi)^{-1/2}, \quad r \to 0, \tag{2.58}$$

where v is a smooth and bounded function on $(0,\alpha)$.

The singularity index τ in Ref. (2.58) is determined by the value of angle α. However, τ is calculated only approximately by numerical methods, for example, as a solution to transcendental equation (4.1) in Ref. [88]. Two limiting cases are known for τ [88]:

$$\lim_{\alpha\to 2\pi-0} \tau(\alpha) = 0 \quad \text{and} \quad \lim_{\alpha\to+0} \tau(\alpha) = 1,$$

so that $\tau(\alpha) \sim 1 - (\ln \frac{1}{\alpha})^{-1}$ for $\alpha \to +0$.

Let us present in the conclusion the table of approximate values of τ calculated in Ref. [88]. The analysis performed in Section 2.6 proves that the singularity index for u in the vicinity of a corner point coincides with $\tau(\alpha)$.

α/π	0.	0.0500	0.1161	0.1250	0.2500	0.3750	0.5000
$\tau(\alpha)$	1.0000	0.8705	0.8350	0.8317	0.7820	0.7384	0.6956
α/π	0.6250	0.7500	0.8750	0.9000	0.9500	1.0000	1.1250
$\tau(\alpha)$	0.6517	0.6057	0.5561	0.5456	0.5243	0.5022	0.4448
α/π	1.2500	1.3750	1.5000	1.6250	1.7500	1.8750	2.0000
$\tau(\alpha)$	0.3799	0.3073	0.2277	0.1444	0.0702	0.0281	0.0000

Let us formulate the statement that completes the study of the solvability of the diffraction problem for a planar screen Ω.

Theorem 4 *For $\Im k \geq 0$ and $k \neq 0$, diffraction problem (2.1)–(2.6) is uniquely solvable for arbitrary E^0 and H^0 that satisfy condition (2.7).*

□ Uniqueness of solution follows from Theorem 1. By virtue of Corollary 1 from Theorem 3 and Statement 9, for every right-hand side $f \in C^\infty(\overline{\Omega})$, there exists a solution u of equation (2.19) that satisfies conditions (2.14) and (2.15). Choosing f in accordance with (2.21), we use formulas (2.11)–(2.13) to obtain the solution of problem (2.1)–(2.6). □

Corollary 2 *For $\Im k \geq 0$ and $k \neq 0$, every solution of problem (2.1)–(2.6) can be represented in the form of the vector potential (2.11)–(2.13), where function u satisfies conditions (2.14) and (2.15).*

Chapter 3

Diffraction by a system of arbitrary bounded screens

In this chapter, we consider the vector problem of diffraction of an electromagnetic field by a system of bounded, perfectly conducting screens of arbitrary form Ω. The main idea of the approach is again in the transition to the analysis of a PDO on Ω. However, in comparison with the case of planar screens, the method becomes substantially different. In place of a usual space of vectors defined on a planar screen, we have to consider a space of tangential vectors defined at every point of the surface, that is, a tangential bundle.

We consider the surface $\overline{\Omega}$ in a natural manner as a submanifold with an edge of a certain enveloping manifold M with a Riemannian metric. Here, the spaces of solutions and images (W and W') consist of the cross sections of vector bundles over Ω. We analyze differential operators and PDOs on Ω using the calculus of the symbols of PDOs that act in the cross sections of vector bundles [35, 40].

Not only the method of the study is changed, but also the results. Subspaces W and W' are not generally orthogonal, and the PDO splitting is not diagonal. For $\Im k \geq 0$ and $k \neq 0$, we prove the PDO injectivity using the unique solvability of the initial problem. Nevertheless, fundamental results concerning the Fredholm property of PDOs and solvability of the diffraction problem remain valid.

In Section 3.1, we define a manifold with an edge $\overline{\Omega}$ (on which the diffraction problem is considered) as a submanifold M. Note that it is not necessary to choose a specific cover of $\overline{\Omega}$ and a specific coordinate system in the vicinity of the edge $\partial\Omega$. Then, we introduce the spaces W and W' of the cross sections of vector bundles over Ω. The integrodifferential equation on Ω will be considered on the pair of these spaces. Using the analysis in local coordinates, we prove statements that describe basic properties of these spaces. Decompositions of spaces W and W' into direct sums of subspaces W_1 and W_2 and W^1 and W^2, respectively, is the most important of these properties. Such a decomposition is necessary for a diagonal splitting of a principal part of a PDO and enables one to investigate its symbols.

In Section 3.2, we consider representations of fields in the form of vector potentials and derive the main integrodifferential equation on the screen. We prove the statements that are vector analogs of the known theorems describing discontinuities of single-layer and double-layer potentials and use these statements to establish the unique solvability of the equation on Ω.

In Section 3.3, the integrodifferential equation is reduced to a vector PDO acting on the manifold $\overline{\Omega}$. We separate the principal part of the corresponding integrodifferential operator in the local coordinates and use the gluing procedure to construct the PDO that differs from the initial operator by an operator with an infinitely smooth kernel. We prove that the PDO is bounded as an operator acting on the pair of spaces $W \to W'$. The integrodifferential equation is considered as a pseudodifferential equation and the generalized solution is defined.

Section 3.4 is a central part of Chapter 3. In this section, we separate the principal part of the PDO and perform its diagonal splitting on subspaces W_1 and W_2, which enables us to prove that, for $k \neq 0$, the PDO: $W \to W'$ is a Fredholm operator with the zero index. Note that again, the main symbol of the PDO is formally degenerate, and we manage to prove that the principal part of the PDO is continuously invertible for $k \neq 0$ only because of an asymmetric structure of space W.

Before to prove the unique solvability of the pseudodifferential equation on Ω in spaces W and W', we elucidate the smoothness of this solution in Ω. Then, we use the uniqueness theorem to show the unique solvability of the equation for $\Im k \geq 0$ and $k \neq 0$. We prove the unique solvability of the initial diffraction problem on Ω and the possibility of representing every solution in the form of a vector potential. In conclusion, we consider the order of singularity of a solution to the pseudodifferential equation in the vicinity of the edge $\partial\Omega$, which is important for various practical applications.

In Section 3.5, we consider the dependence of solutions to the pseudodifferential equation and the initial diffraction problem on parameter k. The principle of limiting absorption is established, as well as the discreteness of the set of scattering frequencies of the diffraction problem situated in the lower half-plane of the complex plane k.

3.1 The spaces W and W' of the cross sections of vector bundles over Ω

Let M be a closed, connected, oriented three-dimensional C^∞-surface (note that a closed surface is a two-dimensional compact manifold without an edge). Let $\overline{\Omega} \subset M$ be a submanifold with an edge of a manifold M, which is not necessarily connected and has a finite number of connected components; each of these components has the dimensionality two. We assume that the edge $\Gamma := \partial\Omega$ is a smooth C^∞-curve without points of selfintersection.

We denote by TM the tangential bundle over M with a standard inner product in the layer T_xM (tangential plane). We fix a finite cover $U = \{U_\alpha\}$ of M by coordinate vicinities and denote by

$$\chi_\alpha : U_\alpha \to V_\alpha \subset \mathrm{R}^2$$

the local maps and by $\{\varphi_\alpha\}$, a partition of unity subordinated to U. For every smooth cross section $u \in C^\infty(M)$ of the bundle TM, we introduce the functions [38]

$$u_\alpha = \varphi_\alpha u \in C_0^\infty(V_\alpha), \quad u_\alpha = (u_\alpha^1, u_\alpha^2),$$

identifying the set U_α with its image in R^2. For a scalar function $g \in C^\infty(M)$, we set $g_\alpha = \varphi_\alpha g$. For every $s \in \mathrm{R}$, we define the Sobolev space $H^s(M)$ as a supplement of $C^\infty(M)$ with respect to the norm $\|\cdot\|_s$, where

$$\|u\|_s^2 = \sum_\alpha (\|u_\alpha^1\|_s^2 + \|u_\alpha^2\|_s^2), \quad \|g\|_s^2 = \sum_\alpha \|g_\alpha\|_s^2.$$

For any other cover, partition of unity, and maps, the norms are equivalent, so that the definition of the space $H^s(M)$ is correct.

For every $s \in \mathrm{R}$, we set, according to the notation used in Ref. [40],

$$H^s(\Omega) := \{u|_\Omega : u \in H^s(M)\}$$

and

$$\tilde{H}^s(\overline{\Omega}) := \left\{u \in H^s(M) : \ supp\, u \subset \overline{\Omega}\right\}.$$

The space $\tilde{H}^s(\Omega)$ may be constructed as a closure of $C_0^\infty(\Omega)$ with respect to the norm $\|\cdot\|_s$. Note that, in the scalar and vector cases, the designation H^s is the same; however, it is always clear which space is considered.

Let us define operations of surface divergence and surface gradient. We assume that the cover U and local maps are chosen so that, in the local coordinates, the first quadratic form of the surface $U = U_\alpha$ can be represented as [36, p. 111] $dl^2 = G(x_1, x_2)(dx_1^2 + dx_2^2)$, and $G = G_\alpha$. For every $U = U_\alpha$, we set

$$div\, u = \frac{1}{G}\left(\frac{\partial(u_1\sqrt{G})}{\partial x_1} + \frac{\partial(u_2\sqrt{G})}{\partial x_2}\right)$$

and

$$grad\, g = \frac{1}{\sqrt{G}}\left(\frac{\partial g}{\partial x_1}e_1 + \frac{\partial g}{\partial x_2}e_2\right).$$

Then, for $u \in C_0^\infty(\Omega)$ and $g \in C_0^\infty(\Omega)$, we have

$$div\, u = \sum_\alpha div\, u_\alpha, \quad grad\, g = \sum_\alpha grad\, g_\alpha,$$

where $u_\alpha = \varphi_\alpha u$ and $g_\alpha = \varphi_\alpha g$.

We define the Hilbert space $W = W(\overline{\Omega})$ as a supplement of $C_0^\infty(\Omega)$ with respect to the norm $\|\cdot\|_W$,

$$\|u\|_W^2 = \|u\|_{-1/2}^2 + \|div\, u\|_{-1/2}^2,$$

so that the inner product is

$$(u, v)_W = (u, v)_{-1/2} + (div\, u,\ div\, v)_{-1/2}.$$

Assume that u belongs to $C_0^\infty(U)$ with $supp\, u \subset U$ and

$$\chi : U \to V \subset \mathrm{R}^2$$

is the corresponding map. Let is find $h \in C_0^\infty(V)$:

$$\begin{cases} \Delta h &= div\, u \quad \text{in} \quad V, \\ h|_{\partial V} &= 0, \end{cases}$$

$$\Delta h := div\, grad\, h = \frac{1}{G}\left(\frac{\partial^2 h}{\partial x_1^2} + \frac{\partial^2 h}{\partial x_2^2}\right).$$

If $b := grad\, h$ and $a := u - b$, then $div\, a = 0$. In the general case, for $u = \sum_\alpha u_\alpha$, we set $h = \sum_\alpha h_\alpha$, $b = \sum_\alpha b_\alpha$, $a = \sum_\alpha a_\alpha$, and $b = grad\, h$.

Consider the projector $Pu := b$, where $u \in C_0^\infty(\Omega)$. Note that $P^2u = Pb = b$, since $div\, u = div\, b$. The following estimate is valid:

$$\|b\|_W^2 = \|b\|_{-1/2}^2 + \|div\, u\|_{-1/2}^2 \leq C\|div\, u\|_{-1/2}^2 \leq C\|u\|_W^2.$$

Hence, P is bounded in W. Let W_i be a a closure of W_i^0 $(i = 1, 2)$ with respect to the norm $\|\cdot\|_W$:

$$W_1 = \overline{W_1^0} \text{ where } W_1^0 := \{u \in C_0^\infty(\Omega) : div\, u = 0\},$$

$$W_2 = \overline{W_2^0} \text{ where } W_2^0 := \{u \in C_0^\infty(\Omega) : u = grad\, h, h \in C_0^\infty(\Omega)\}.$$

Then, $PW = W_2$ and $(1 - P)W = W_1$. Taking into account that P is bounded, one can prove the following statement (see, for example, Ref. [26, p. 198]).

Statement 17 *The space W is expanded into a direct sum of closed subspaces W_1 and W_2:*

$$W = W_1 \oplus W_2.$$

Now, we extend operation div to $u \in H^s(M)$ (with the help of the Fourier transform [35, p. 158]) to obtain

Statement 18 $W = \{u \in \tilde{H}^{-1/2}(\overline{\Omega}) : div\, u \in \tilde{H}^{-1/2}(\overline{\Omega})\}$.

Statement 19 *The embeddings*

$$\tilde{H}^{1/2}(\overline{\Omega}) \subset W \subset \tilde{H}^{-1/2}(\overline{\Omega})$$

are continuous, and the norms can be estimated as

$$\|u\|_{-1/2} \leq \|u\|_W \leq C_0\|u\|_{1/2}.$$

In addition to this,

$$\|u\|_W = \|u\|_{-1/2}$$

for $u \in W_1$ and

$$C_1\|u\|_{1/2} \leq \|u\|_W \leq C_2\|u\|_{1/2}$$

for $u \in W_2$.

□ The proofs of two latter statements are similar to the corresponding proofs in Chapter 2. It is necessary to repeat considerations for every U_α and then, to obtain the results. Only the last inequalities should be verified separately. Let us represent operator P in the form

$$Pu = grad\, \Delta^{-1}(div\, u).$$

For $u \in W_2$, we have the estimates

$$\begin{aligned} \|u\|_{1/2} &= \|Pu\|_{1/2} \leq C\|\Delta^{-1}(div\, u)\|_{3/2} \\ &\leq C_2\|div\, u\|_{-1/2} \leq C_2\|u\|_W. \end{aligned}$$

On the other hand,

$$\begin{aligned} \|u\|_W^2 &= \|div\, u\|_{-1/2}^2 + \|u\|_{-1/2}^2 \\ &\leq C\|u\|_{1/2}^2 + \|u\|_{-1/2}^2 \leq C_1^2\|u\|_{1/2}^2, \end{aligned}$$

which completes the proof of inequalities. □

Let us introduce the operations that are defined in the vicinity of $U = U_\alpha$, by the formulas

$$rot_\nu u = \frac{1}{G}\left(\frac{\partial(u_2\sqrt{G})}{\partial x_1} - \frac{\partial(u_1\sqrt{G})}{\partial x_2}\right)$$

and

$$grad' g = \frac{1}{G}\left(\frac{\partial g}{\partial x_2}e_1 - \frac{\partial g}{\partial x_1}e_2\right)$$

in the local coordinates. In the general case, for $u = \sum\limits_\alpha u_\alpha$ and $g = \sum\limits_\alpha g_\alpha$,

$$rot_\nu u = \sum_\alpha rot_\nu u_\alpha$$

and

$$grad' g = \sum_\alpha grad' g_\alpha.$$

Consider in more detail the expansion $u = u_1 + u_2$, where $u_1 \in W_1$ and $u_2 \in W_2$. Every element $u \in C_0^\infty(\Omega)$ can be represented as

$$u = (1-P)u + Pu = grad' g + grad\, h =$$

$$grad' \Delta^{-1}(rot_\nu u) + grad\, \Delta^{-1}(div\, u),$$

where g is obtained as a result of solving the problem

$$\begin{cases} \Delta g_\alpha & = & rot_\nu u_\alpha \text{ in } V_\alpha, \\ g_\alpha|_{\partial V_\alpha} & = & 0, \end{cases}$$

$$g = \sum_\alpha g_\alpha.$$

Thus, $(1-P)u = grad'\Delta^{-1}(rot_\nu u)$. We define the action of operators P and $1-P$ on the whole space W using the continuity. Obviously,

$$div\, grad' g \equiv 0 \quad \text{and} \quad rot_\nu\, grad\, h \equiv 0.$$

Below, we will use two relationships that follow from the general Stokes formula [36]:

$$\int_M u \cdot grad\, b\, ds = -\int_M (div\, u) b\, ds \tag{3.1}$$

and

$$\int_M u \cdot grad' b\, ds = -\int_M (rot_\nu u) b\, ds, \tag{3.2}$$

$$u, b \in C^\infty(M).$$

Formulas (3.1) and (3.2) can be obtained if to transfer to local coordinates.

The next statements describe some properties of the space $W' = (W(\overline{\Omega}))'$ which is antidual with respect to W.

Statement 20 *The space*

$$W' = \left\{ f|_\Omega : f \in H^{-1/2}(M),\ rot_\nu f \in H^{-1/2}(M) \right\}$$

and

$$H^{1/2}(\Omega) \subset W' \subset H^{-1/2}(\Omega).$$

□ Let us denote by $'W(M)$ the space of elements from $H^{-1/2}(M)$ with a finite norm

$$\|f\|'^2 := \|rot_\nu f\|^2_{-1/2} + \|f\|^2_{-1/2}.$$

Consider the space $W(M)$ defined as a supplement of $C^\infty(M)$ with respect to the norm

$$\|u\|^2_W = \|u\|^2_{-1/2} + \|div\, u\|^2_{-1/2}$$

and the corresponding antidual space $W'(M)$.

According to the Riesz theorem, the general form of an antilinear continuous functional on $W(M)$ is $(v, u)_W$ for a certain $v \in W(M)$, an the norm of this functional equals $\|v\|_W$. Then, since $C^\infty(M)$ is dense in $W(M)$, the functionals $(v_0, u)_W$, $v_0 \in C^\infty(M)$, are dense in $W'(M)$. Using local coordinates, it is easy to check that if $v \in C^\infty(M)$, then there exists $f \in C^\infty(M)$ such that

$$(v, u)_W = \int_M f \cdot \overline{u}\, ds \equiv (f, u), \quad \forall u \in C^\infty(M).$$

Usually, this functional is identified with element f. In this sense, one can say that $C^\infty(M)$ is dense in $W'(M)$.

Applying formula (3.2), we obtain the following estimates:

$$\begin{aligned}
|(f,u)| &= |(f, grad' g) + (f, grad\, h)| \\
&= |-(rot_\nu f, g) + (f, grad\, h)| \\
&\le \|g\|_{1/2}\|rot_\nu f\|_{-1/2} + \|grad\, h\|_{1/2}\|f\|_{-1/2} \\
&\le C\left(\|u\|_{-1/2}\|rot_\nu f\|_{-1/2} + \|div\, u\|_{-1/2}\|f\|_{-1/2}\right) \\
&\le C\|u\|_W\|f\|', \\
&\qquad u, f \in C^\infty(M).
\end{aligned}$$

Therefore,

$$\|f\|_{W'} = \sup_{0 \ne u \in W} \frac{|(f,u)|}{\|u\|_W} \le C\|f\|'.$$

On the other hand, setting

$$u_0 = f = grad'\Delta^{-1}(rot_\nu f) + grad\Delta^{-1}(div\, f)$$

and again applying formulas (3.1) and (3.2), we have

$$\begin{aligned}
(f, u_0) &= \|f\|^2 \\
&= -(rot_\nu f, \Delta^{-1} rot_\nu f) - (div\, f, \Delta^{-1} div\, f) \\
&\ge C\left(\|rot_\nu f\|^2_{-1/2} + \|div\, f\|^2_{-1/2}\right)
\end{aligned}$$

and

$$\|f\|^2 \geq \|f\|^2_{-1/2}.$$

Consequently,

$$\begin{aligned}\|f\|_{W'} &\geq \frac{|(f,u_0)|}{\|u_0\|_W}\\ &\geq \frac{C_0(\|rot_\nu f\|^2_{-1/2} + \|div\, f\|^2_{-1/2} + 2\|f\|^2_{-1/2})}{\|f\|_W}\\ &\geq 2C_0\|f\|'.\end{aligned}$$

The above estimates prove that norms $\|\cdot\|_{W'(M)}$ and $\|\cdot\|'$ are equivalent and spaces $W'(M)$ and $'W(M$ are antidual. Finally, taking into account that $C_0^\infty(\Omega)$ is dense in $W(\overline{\Omega})$, we find that the space of restrictions of the elements from $'W(M)$ on Ω is antidual with respect to $W(\Omega)$. The antiduality relationships and Statement 8 yield the continuous embeddings

$$H^{1/2}(\Omega) \subset W' \subset H^{-1/2}(\Omega)$$

□ We note that, from the proof of the latter statement, it follows that $C^\infty(\overline{\Omega})$ is dense in W'.

Statement 21 *The space W' is expanded into the direct sum of closed subspaces,*

$$W' = W^1 \oplus W^2,$$

where

$$W^1 := \{f \in W' \, : \, div\, f = 0\},$$
$$W^2 := \{f \in W' \, : \, rot_\nu\, f = 0\}.$$

□ Let $q : W'(\Omega) \to W'(M)$ is an operator of continuation of an element from Ω on M and $p : W'(M) \to W'(\Omega)$ is an operator of restricting an element from M on Ω; p and q are continuous operators. For every $\tilde{f} \in W'(M)$ (it is sufficient to consider $\tilde{f} \in C^\infty(M)$), one can prove the validity of the expansion

$$\tilde{f} = grad'\Delta^{-1}(rot_\nu \tilde{f}) + grad\Delta^{-1}(div\tilde{f}) = (1-\tilde{P})\tilde{f} + \tilde{P}\tilde{f},$$

where $\tilde{P} : W'(M) \to W'(M)$ is a bounded projector. Then, assuming that $f \in W'(\Omega)$, we have

$$f = p(qf) = p\tilde{f} = p(1-\tilde{P})\tilde{f} + p\tilde{P}\tilde{f} = f_1 + f_2,$$

so that

$$div\, f_1 = div\, p(1-\tilde{P})\tilde{f} = p\, div(1-\tilde{P})\tilde{f} = 0$$

and

$$rot_\nu f_2 = rot_\nu p\tilde{P}\tilde{f} = p\, rot_\nu \tilde{P}\tilde{f} = 0.$$

Here, we have taken into account that

$$supp\, div\, f \subset supp\, f$$

and

$$supp\, rot_\nu f \subset supp\, f.$$

Spaces W^1 and W^2 are closed because operators p, q, and $\tilde{P}$ are bounded. □

All results of this section remain valid if to consider a manifold M without an edge in place of a submanifold $\overline{\Omega}$ with an edge.

3.2 Representation of solutions and the system of integrodifferential equations on screens

By virtue of the definition given in Section 3.1,

$$\Omega = \bigcup_j \Omega_j, \quad \overline{\Omega}_i \bigcap \overline{\Omega}_j = \emptyset \quad (i \neq j)$$

is a union of a finite number of connected, oriented, unclosed nonintersecting C^∞-surfaces in R^3. The edge $\partial\Omega_j = \overline{\Omega}_j \setminus \Omega_j$ of the surface Ω_j is a smooth C^∞-curve without points of self-intersection, and $\Gamma = \partial\Omega = \bigcup_j \partial\Omega_j$.

The problem of diffraction of a monochromatic electromagnetic field by a system of bounded screens Ω does not differ from that considered in Section 2.1, if to assume that Ω is not a planar screen but a system of arbitrarily shaped screens described above. Thus, one can consider problem (2.1)–(2.6) with conditions (2.7) and (2.8) for an incident field, replacing (2.1) by a similar condition

$$E, H \in C^2\left(R^3 \setminus \overline{\Omega}\right) \bigcap_{\delta>0} C\left(\overline{M_+} \setminus \Gamma_\delta\right) \bigcap_{\delta>0} C\left(\overline{M_-} \setminus \Gamma_\delta\right)$$

for arbitrary screens, where M_+ and M_- are, respectively, the external and internal parts of surface M. By ν we will denote the external unit normal to M.

The uniqueness theorem for the diffraction problem on Ω remains valid and its proof literally coincides with the proof presented in Section 2.1. Therefore, when a system of screens Ω will be considered, we will also refer to this theorem.

We will look for the solution E, H to problem (2.1)–(2.6), which again will be called quasiclassical, in the form of a vector potential

$$E = ik^{-1}\left(Grad\, Div\,(A_1 u) + k^2 A_1 u\right), \tag{3.3}$$

$$H = Rot\,(A_1 u), \tag{3.4}$$

$$A_1 u = \frac{1}{4\pi}\int_\Omega \frac{e^{ik|x-y|}}{|x-y|} u(y)\, ds; \quad x = (x_1, x_2, x_3) \notin \overline{\Omega}. \tag{3.5}$$

Here, $u(y)$ is a tangential vector field defined on Ω. We have the relationship $u(y)\cdot\nu(y) = 0$ for all $y \in \Omega$, where $\nu(y)$ is the unit normal vector to Ω at a point y, so that u is the surface current density on Ω.

Sometimes, it will be convenient to consider not the tangential field u on Ω, but its continuation $U(x)$ to the domain $G_t \subset R^3$ $(\Omega \subset G_t)$, which is defined as

$$G_t := \{x : x = y + t\nu(y),\ y \in \Omega, \|t| < \delta\}.$$

We choose parameter δ so that $y \in \Omega$ would be the only closest point of Ω for $x = y \pm \delta\nu(y)$. Obviously, such a value of δ always exists (since $\Omega \in C^\infty$). Thus, G_t is a normal tubular vicinity of Ω, if to consider Ω as a correct manifold with an edge, embedded to a certain enveloping manifold that contains G_t [63].

We set $U(x) = u(y)$ and $U_1(x) = \nu(y) \times u(y)$ for all $x = y + t\nu(y)$ with $|t| < \delta$. Then, $u(y) = U_1(y) \times \nu(y)$, where $y \in \Omega$. Note that δ is chosen so that this definition of the vector field $U(x)$ is correct. For a smooth tangential field, we define the divergence

$$div\, u(y) := (Div\, U)|_\Omega, \quad y \in \Omega.$$

In the Cartesian coordinate system (z_1, z_2, z_3) with the unit vectors $e_1, e_2, e_3 = \nu(y)$ and the origin at the point y,

$$(Div\, U)|_{\Omega} = \frac{\partial U^1}{\partial z_1} + \frac{\partial U^2}{\partial z_2}.$$

We will assume that u satisfies the conditions

$$u \in W(\overline{\Omega}), \tag{3.6}$$

$$u,\ div\, u \in C^1(\Omega). \tag{3.7}$$

The choice of condition (3.7), which is similar to (2.15), is explained in Chapter 1.

Proceeding to local coordinates, it is easy to prove (see Section 2.3) that

$$A_1 u \in C^{\infty}(\mathrm{R}^3 \setminus \overline{\Omega}).$$

As has been shown in Ref. [75], the operator

$$A_1 : \tilde{H}^{-1/2}(\overline{\Omega}) \to H^1_{loc}(\mathrm{R}^3)$$

acts continuously on this pair of spaces.

Let us prove that

$$Div\,(A_1 u) = A_1(div\, u), \quad \mathrm{x} \notin \overline{\Omega}, \quad u \in W.$$

It is sufficient to prove this equality for functions $u \in C_0^{\infty}(\Omega)$, because this set is dense in W and in $\tilde{H}^{-1/2}(\overline{\Omega})$. In this case, we have $A_1(div\, u) \in H^1_{loc}(\mathrm{R}^3)$.

Let $u \in C_0^{\infty}(\Omega)$. Calculating the divergence under the integral sign and using the Stokes formula, we obtain

$$\begin{aligned}
Div\,(A_1 u) &= \frac{1}{4\pi}\int_{\Omega} u(y)\cdot Grad_{\mathrm{x}}\left(\frac{e^{ik|\mathrm{x}-y|}}{|\mathrm{x}-y|}\right) ds \\
&= -\frac{1}{4\pi}\int_{\Omega} U(y)\cdot Grad_{y}\left(\frac{e^{ik|\mathrm{x}-y|}}{|\mathrm{x}-y|}\right) ds \\
&= -\frac{1}{4\pi}\int_{\Omega} Div_{y}\left(\frac{e^{ik|\mathrm{x}-y|}}{|\mathrm{x}-y|} U_1(y)\times\nu(y)\right) ds \\
&\quad +\frac{1}{4\pi}\int_{\Omega} \frac{e^{ik|\mathrm{x}-y|}}{|\mathrm{x}-y|} Div\, U(y)\, ds \\
&= -\frac{1}{4\pi}\int_{\Omega} \nu(y)\cdot Rot_{y}\left(\frac{e^{ik|\mathrm{x}-y|}}{|\mathrm{x}-y|} U_1(y)\right) ds \\
&\quad +\frac{1}{4\pi}\int_{\Omega} \frac{e^{ik|\mathrm{x}-y|}}{|\mathrm{x}-y|} div\, u(y)\, ds \\
&= -\frac{1}{4\pi}\oint_{\Gamma} \frac{e^{ik|\mathrm{x}-y|}}{|\mathrm{x}-y|} U_1(y)\cdot t(y)\, dl \\
&\quad +\frac{1}{4\pi}\int_{\Omega} \frac{e^{ik|\mathrm{x}-y|}}{|\mathrm{x}-y|} div\, u(y)\, ds \\
&= \frac{1}{4\pi}\int_{\Omega} \frac{e^{ik|\mathrm{x}-y|}}{|\mathrm{x}-y|} div\, u(y)\, ds.
\end{aligned}$$

Here, $t(y)$ is the tangential unit vector of Γ. Note that we have used the relationship $Rot\,\nu(\mathrm{x}) = 0$, which is valid under the assumption that $\nu(\mathrm{x})$ is continued to G_t: $\nu(\mathrm{x}) = \nu(y)$ for $\mathrm{x} = y + \nu(y)t$, $|t| < \delta$. If u satisfies conditions (3.6) and (3.7), then (3.3) is equivalent to

$$E = ik^{-1}\left(Grad\,A_1(div\,u) + k^2 A_1 u\right), \quad \mathrm{x} \in \mathrm{R}^3 \setminus \overline{\Omega}. \tag{3.8}$$

The fields $E, H \in C^\infty(\mathrm{R}^3 \setminus \overline{\Omega})$, defined by formulas (3.3) and (3.4) or (3.8), satisfy the Maxwell equations (2.2) in $\mathrm{R}^3 \setminus \overline{\Omega}$ and conditions at infinity (2.5) and (2.6) or (2.9). One can verify these statements directly using properties of the kernels in integral representation (3.5) and of the functions E and H chosen in the form (3.3) and (3.4).

The statements that are proved for planar screens remain valid for the limiting values of E and H determined when a point x tends to surface Ω. Let $x \in \Omega$. Then, the tangential components of the field E and the normal component of the field H are continuous up to Ω (excluding the edge points on Γ). More precisely,

$$\lim_{t\to 0} \nu(x) \times E(x + \nu(x)t) = \nu(x) \times E^{(0)}(x),$$

$$\lim_{t\to 0} \nu(x) \cdot H(x + \nu(x)t) = \nu(x) \cdot H^{(0)}(x), \quad x \in \Omega.$$

For the normal component of E and the tangential components of H, we obtain the formulas

$$\lim_{t\to\pm 0} \nu(x) \cdot E(x + \nu(x)t) = \mp\frac{i}{2k} div\,u(x) + \nu(x) \cdot E^{(0)}(x), \tag{3.9}$$

$$\lim_{t\to\pm 0} \nu(x) \times H(x + \nu(x)t) = \pm\frac{1}{2}u(x) + \nu(x) \times H^{(0)}(x), \quad x \in \Omega, \tag{3.10}$$

where

$$E^{(0)}(x) = \frac{i}{4\pi k}\left(\int_\Omega Grad_{\mathrm{x}}\left(\frac{e^{ik|\mathrm{x}-y|}}{|\mathrm{x}-y|} div\,u\right)_{\mathrm{x}=x} ds + k^2 \int_\Omega \frac{e^{ik|x-y|}}{|x-y|} u\,ds\right),$$

$$H^{(0)}(x) = \frac{1}{4\pi} \int_\Omega Rot_{\mathrm{x}}\left(\frac{e^{ik|\mathrm{x}-y|}}{|\mathrm{x}-y|} u\right)_{\mathrm{x}=x} ds,$$

where $u = u(y)$ and singular integrals are understood in the sense of valeur principal.

In order to prove the latter expressions, we fix $x \in \Omega$ and the vicinity $O_\varepsilon(x) = \{y \in \Omega : |y - x| < \varepsilon\}$. Consider the integral

$$\begin{aligned} A_1 u &= A_1' u + A_1'' \\ &= \frac{1}{4\pi}\int_\Omega \varphi(|x-y|)\frac{e^{ik|\mathrm{x}-y|}}{|\mathrm{x}-y|} u\,ds + \frac{1}{4\pi}\int_\Omega (1 - \varphi(|x-y|))\frac{e^{ik|\mathrm{x}-y|}}{|\mathrm{x}-y|} u\,ds, \\ u &= u(y), \quad \mathrm{x} = x + \nu(x)t, \end{aligned}$$

for $t \to 0$, where $\varphi(r) \in C^\infty(\overline{\mathrm{R}^1_+})$, $0 \le \varphi \le 1$, $\varphi \equiv 1$ for $r < \varepsilon/2$, and $\varphi \equiv 0$ for $r \ge \varepsilon$. Choose ε so that for $O_\varepsilon(x) \subset \Omega$, $dist(x, \Gamma) > \varepsilon$. Then, $(A_1'' u)(x) \in C^\infty(M)$, because u is a finite function and the kernel in $A_1'' u$ is infinitely differentiable. Performing differentiation and transition to a limit in $A_1''u$ under the integral sign, we obtain

$$\begin{aligned} A_1'u &= \frac{1}{4\pi}\int\limits_{O_\epsilon(x)} \varphi(|x-y|)\frac{e^{ik|\mathrm{x}-y|}}{|\mathrm{x}-y|}u\,ds \\ &= \frac{1}{4\pi}\int\limits_{O_{\epsilon/2}(x)} \frac{e^{ik|\mathrm{x}-y|}}{|\mathrm{x}-y|}u\,ds + \frac{1}{4\pi}\int\limits_{O_\epsilon(x)\setminus O_{\epsilon/2}(x)} \frac{e^{ik|\mathrm{x}-y|}}{|\mathrm{x}-y|}\varphi(|x-y|)u\,ds \\ &= I_1 + I_2, \quad u = u(y). \end{aligned}$$

Since the kernel of I_2 is not singular, we can differentiate I_2 with respect to x_j and calculate the limit under the integral sign (here, u is a usual differentiable function). I_1 is a weakly singular integral, and the limit for $t \to 0$ can be also calculated under the integral sign.

Now, let us consider the integral

$$I = I(\mathrm{x}) = \int\limits_{\Omega_\epsilon} \frac{e^{ik|\mathrm{x}-y|}}{|\mathrm{x}-y|}g(y)\,ds, \quad \Omega_\varepsilon := O_{\varepsilon/2}(x),$$

where $g(y)$ is a smooth scalar function on $\overline{\Omega}_\varepsilon$, and the derivative $\partial I/\partial l$ along the (fixed) direction l at the point $\mathrm{x} = x + \nu(x)t$, $t \neq 0$. Performing the same transformation as in Section 2.3, we separate in $\partial I/\partial l$ a part that contains the principal singularity:

$$\frac{\partial I}{\partial l} = g(x)J_1 + J_2,$$

where

$$J_1 = \int\limits_{\Omega_\epsilon} \frac{e^{ik|\mathrm{x}-y|}(ik|\mathrm{x}-y|-1)}{|\mathrm{x}-y|^3} l\cdot(\mathrm{x}-y)\,ds. \tag{3.11}$$

It is sufficient to analyze J_1, since J_2 is a weakly singular integral, in which transition to a limit can be preformed under the integral sign.

Let us proceed in (3.11) to new Cartesian coordinates z_1, z_2, z_3 with the unit vectors $e_1, e_2, e_3 = \nu(x)$ and the origin at the point x. Then, integral J_1 can be represented as

$$J_1 = \int\limits_{\Omega'_\epsilon} \frac{e^{ik|z-z^0|}(ik|z-z^0|-1)}{|z-z^0|^3} l\cdot(z^0-z)\sqrt{1+\left(\frac{\partial f}{\partial z_1}\right)^2+\left(\frac{\partial f}{\partial z_2}\right)^2}\,dV.$$

Here, $dV = dz_1dz_2$, $z^0 = (0,0,z_3^0)$, $z = (z_1, z_2, f(z_1,z_2))$, the function $z_3 = f(z_1, z_2)$ describes the surface Ω_ε in the vicinity of the origin, and Ω'_ε is the projection of Ω_ε on the tangential plane Oz_1z_2. We obtain the asymptotic relations

$$\sqrt{1+\left(\frac{\partial f}{\partial z_1}\right)^2+\left(\frac{\partial f}{\partial z_2}\right)^2} = 1 + O(|z'|^2),$$

$$f(z') = O(|z'|^2), \quad |z'| \to 0, \quad z' = (z_1, z_2).$$

Then, $J_1 = J_1' + J_2'$, where

$$J_1' = \int\limits_{\Omega_\epsilon^0} \frac{e^{ik|z'-z^0|}(ik|z'-z^0|-1)}{|z'-z^0|^3} l\cdot(z^0-z')\,dz',$$

$\Omega_\varepsilon^0 = \{z' : |z| < \varepsilon_0\} \subset \Omega'_\varepsilon$ is a certain subdomain of Ω'_ε, and the term J'_2 corresponds to integrals that do not contain the principal singularity and in which one can perform transition to a limit under the integral sign.

Integral J'_1 was considered in Section 2.2, where it was shown that $J'_1 = 0$ if $l \cdot \nu(x) = 0$ and $\lim_{t\to\pm 0} J'_1 = \mp 2\pi$ if $l = \nu(x)$. Thus, the limiting value of $\partial I/\partial l$ for $t \to \pm 0$ has a break $\mp 2\pi g(x)(l \cdot \nu(x))$, $x \in \Omega$. which yields relationship (3.9), if we set $g = div\, u$. Representing H with the help of vector identities [27] as

$$H = Rot_{\mathrm{x}}(A_1 u) = Grad_{\mathrm{x}} A_1(\nu \cdot u) - \frac{\partial}{\partial \nu}(A_1 u), \tag{3.12}$$

we find that the first term in (3.12) has no break, because $\nu(x) \cdot u(x) = 0$, and the break of the second term equals $\pm u(x)/2$. The proof of formula (3.10) is completed.

From formula (3.10), we obtain

$$u(x) = [\nu(x) \times H(\mathrm{x})]_\Omega, \quad x \in \Omega, \tag{3.13}$$

where $[\,\cdot\,]_\Omega$ denotes the difference of limiting values that are calculated as $t \to +0$ and $t \to -0$ for $\mathrm{x} = x + \nu(x)t$ at a point $x \in \Omega$. This formula explains the physical sense of u.

Let us complete the definitions of the tangential components of H and the normal component of E on each side of Ω by formulas (3.9) and (3.10). Then, E and H will be continuous (on each side) at points of Ω and condition (2.1) will be satisfied. Boundary condition (2.3) yields an integrodifferential equation for u. For x on Ω, from (2.6) and (3.8), we obtain

$$grad_\tau\, A(div\, u) + k^2 A_\tau u = f, \quad x \in \Omega, \tag{3.14}$$

$$Au = \int_\Omega \frac{e^{ik|x-y|}}{|x-y|} u(y)\, ds, \quad A_\tau u := (Au)_\tau, \tag{3.15}$$

$$f = 4\pi i k E^0_\tau\Big|_\Omega, \quad f \in C^\infty(\overline{\Omega}). \tag{3.16}$$

Here,

$$grad_\tau := Grad_{\mathrm{x}} - \nu(x)(\nu(x) \cdot Grad_{\mathrm{x}}), \tag{3.17}$$

$$(Au)_\tau(x) := (Au)(x) - \nu(x)\,(\nu(x) \cdot (Au)(x)), \tag{3.18}$$

$$E^0_\tau(x) := E^0(x) - \nu(x)\left(\nu(x) \cdot E^0(x)\right). \tag{3.19}$$

In the case of a planar screen, equation (3.14) is transformed to (2.19).

As has been shown above,

$$A_1 u \in H^1_{loc}(\mathrm{R}^3), \quad u \in \tilde{H}^{-1/2}(\overline{\Omega}). \tag{3.20}$$

Then, from (3.4), (3.8), and (3.5), we find that $E, H \in L^2_{loc}(\mathrm{R}^3)$ for $u \in W$, which means that condition (2.4) is also valid (the field energy is finite in every bounded spatial volume).

Thus, if u is a solution to (3.14) and satisfies conditions (3.6) and (3.7), formulas (3.3)–(3.5) or (3.8) give a quasiclassical solution to problem (2.1)–(2.6) on Ω. In addition to this, if u is a nontrivial solution, then, by virtue of (3.13), E, H is also a nontrivial solution to (2.1)–(2.6). Finally, applying the uniqueness theorem, we prove the following result.

Theorem 5 *Equation (3.14) has no more than one solution that satisfies conditions (3.6) and (3.7).*

□ The proof is similar to that of Theorem 2 from the previous section. □

In the subsequent sections, we will prove that, for $\Im k \geq 0$ and $k \neq 0$, equation (3.14) is always solvable. Therefore, formulas (3.3)–(3.5) and (3.8) define the unique solution to problem (2.1)–(2.6), and each solution to this problem can be represented as a vector potential.

3.3 Reduction of the problem to the vector pseudodifferential equation on Ω

Consider the kernel of integral operator (3.15). Fix α, $U = U_\alpha$, and $V = V_\alpha$. Let $\chi^{-1} : V \to U$, $\mathrm{x} = \chi^{-1}(x) \in U$, $x = (x_1, x_2) \in V$ be local coordinates on U, where $\mathrm{x} = (\mathrm{x}_1, \mathrm{x}_2, \mathrm{x}_3)$. We have

$$|\mathrm{x} - y| = |\chi^{-1}(x) - \chi^{-1}(x_0)| = \Phi(x, x_0)|x - x_0|,$$

where $\Phi(x, x_0) \in C^\infty(V \times V)$ (since U is a C^∞-surface) and $\Phi(x, x_0) > 0$ for $x, x_0 \in V$; in addition to this, $\Phi(x, x_0) = \Phi(x_0, x)$. We set

$$\Phi(x, x_0) = \Theta(x_0) + \Pi(x, x_0), \quad \Theta(x_0) = \Phi(x_0, x_0), \quad \Pi \in C^\infty(V \times V),$$

where $\Pi(x_0, x_0) = 0$ for $x_0 \in V$. Then

$$\begin{aligned}
\frac{e^{ik|\mathrm{x}-y|}}{|\mathrm{x} - y|} &= \frac{e^{ik\Phi(x,x_0)|x-x_0|}}{\Phi(x, x_0)|x - x_0|} \\
&= \frac{e^{-|x-x_0|}}{\Phi(x, x_0)|x - x_0|} + \frac{\cos(k\Phi(x, x_0)|x - x_0|) - \operatorname{ch}|x - x_0|}{\Phi(x, x_0)|x - x_0|} \\
&\quad + \frac{i\sin(k\Phi(x, x_0)|x - x_0|) + \operatorname{sh}|x - x_0|}{\Phi(x, x_0)|x - x_0|} \\
&= \frac{e^{-|x-x_0|}}{\Phi(x, x_0)|x - x_0|} + |x - x_0|\Phi_1(x, x_0) + \Phi_2(x, x_0) \\
&= \frac{e^{-|x-x_0|}}{\Theta(x_0)|x - x_0|} + e^{-|x-x_0|}\frac{\Pi_1(x, x_0)}{|x - x_0|} \\
&\quad + |x - x_0|\Phi_1(x, x_0) + \Phi_2(x, x_0),
\end{aligned}$$

where $\Pi_1, \Phi_1, \Phi_2 \in C^\infty(V \times V)$, and $\Pi_1(x_0, x_0) = 0$ for $x_0 \in V$.

On the surface U, the area differential can be represented as

$$ds = G(x)\,dx, \; G(x) = G(x_0) + Q(x, x_0), \; G \in C^\infty(V), \; Q \in C^\infty(V \times V),$$

where $Q(x, x_0) = G(x) - G(x_0)$ and $Q(x_0, x_0) = 0$ for $x_0 \in V$.

If $\Pi_0(x, x_0) \in C^\infty(V \times V)$ and $\Pi_0(x_0, x_0) = 0$ for $x_0 \in V$, then

$$\begin{aligned}
\frac{\Pi_0(x, x_0)}{|x - x_0|} &= \frac{B_0(x_0) \cdot (x - x_0)}{|x - x_0|} + |x - x_0| \left(\widehat{F}_0(x_0) \frac{x - x_0}{|x - x_0|} \right) \cdot \frac{x - x_0}{|x - x_0|} \\
&\quad + |x - x_0|\Phi_0(x, x_0),
\end{aligned}$$

where vector $B_0 \in C^\infty(V)$, matrix $\widehat{F}_0 \in C^\infty(V)$, and function $\Phi_0 \in C^\infty(V \times V)$. The resulting expression takes the form

$$\begin{aligned}\frac{e^{ik\Phi(x,x_0)|x-x_0|}}{\Phi(x,x_0)|x-x_0|}G(x) &= \frac{G(x_0)e^{-|x-x_0|}}{\Theta(x_0)|x-x_0|} + \frac{B(x_0)\cdot(x-x_0)}{|x-x_0|}e^{-|x-x_0|} \\ &+ e^{-|x-x_0|}|x-x_0|\left(\widehat{F}(x_0)\frac{x-x_0}{|x-x_0|}\right)\cdot\frac{x-x_0}{|x-x_0|} \\ &+ |x-x_0|\tilde{\Phi}_1(x,x_0) + \tilde{\Phi}_2(x,x_0), \end{aligned} \tag{3.21}$$

where $G, \Theta \in C^\infty(V)$, $G(x_0) > 0$ and $\Theta(x_0) > 0$ for $x_0 \in V$, $B, \widehat{F} \in C^\infty(V)$, and functions $\tilde{\Phi}_1, \tilde{\Phi}_2 \in C^\infty(V \times V)$.

Depending on the situation, we will consider operators A (that act in a functional space) and A_τ (that act in the cross sections of vector bundles) as PDOs on the manifolds M or Ω. For every coordinate vicinity $U = U_\alpha$ and, respectively, $V = V_\alpha$, we define the restriction of A and A_τ on V by the formulas

$$A_V = p_V A q_V : C_0^\infty(V) \to C^\infty(V),$$
$$\widehat{A}_V = p_V A_\tau q_V : C_0^\infty(V) \to C^\infty(V),$$

where $q_V : C_0^\infty(V) \to C^\infty(M)$ is a natural inclusion (the zero continuation outside V) and $p_V : C^\infty(M) \to C^\infty(V)$ is the operator of restriction that transforms f to $f|_V$. We do not distinguish between operators q_V and p_V and spaces C_0^∞ and C^∞ in the scalar and vector cases. If A_V and $\widehat{A}_V$ are transformed into scalar and matrix PDOs, then A and A_τ are PDOs on the manifolds M [15] or Ω, if to consider the restrictions of A and A_τ on Ω:

$$A : C_0^\infty(\Omega) \to C^\infty(\Omega), \quad A_\tau : C_0^\infty(\Omega) \to C^\infty(\Omega).$$

Let us define the action of A on $C_0^\infty(\Omega)$. Since it is necessary to know only the values Au at points $\mathrm{x} \in \Omega$ (or $\mathrm{x} \in M$), we will use the following representation for A_V:

$$\begin{aligned} A_V u &= \int_V \frac{a_0(x_0)}{|x-x_0|}e^{-|x-x_0|}u(x)\,dx + \int_V e^{-|x-x_0|}\frac{B(x_0)\cdot(x-x_0)}{|x-x_0|}u(x)\,dx \\ &+ \int_V e^{-|x-x_0|}|x-x_0|\left(\widehat{F}(x_0)\frac{x-x_0}{|x-x_0|}\right)\cdot\frac{x-x_0}{|x-x_0|}u(x)\,dx \\ &+ \int_V \eta(|x-x_0|)|x-x_0|\Psi_1(x_0, x-x_0)u(x)\,dx \\ &+ \int_V \eta(|x-x_0|)\Psi_2(x_0, x-x_0)u(x)\,dx. \end{aligned} \tag{3.22}$$

Here,

$$a_0(x_0) = \frac{G(x_0)}{\Theta(x_0)} = \sqrt{G(x_0)},$$

$\eta(t) = 1$ for $t \le R$ and $\eta(t) = 0$ for $t \ge 2R$ is an infinitely differentiable cut function, and R is chosen so large that $\eta(|x-x_0|) \equiv 1$ for $x, x_0 \in V$. In addition to this, in (3.22), $\Psi_i(x_0, x-x_0) = \tilde{\Phi}_i(x, x_0)$, $x, x_0 \in V$ and functions $\Psi_i \in C^\infty(V \times \mathrm{R}^2)$ are smoothly continued on R^2 with respect to the second argument. Formulas (3.15) and (3.22) obviously define one and the same operator for $x, x_0 \in V$, and its definition does not depend on the choice of function η.

Each term in (3.22) is an integral, convolution-type operator. Let us calculate the Fourier transform of the kernels of the first three operators:

$$F\left(\frac{e^{-|x|}}{|x|}\right) = \frac{1}{\langle\xi\rangle},$$

$$F\left(\frac{x}{|x|}e^{-|x|}\right) = -i\frac{\xi}{\langle\xi\rangle^3}, \tag{3.23}$$

$$F\left(e^{-|x|}|x|\left(\widehat{F}(x_0)\frac{x}{|x|}\right)\cdot\frac{x}{|x|}\right) = \frac{tr\widehat{F}(x_0)\langle\xi\rangle^2 - 3(\widehat{F}(x_0)\xi)\cdot\xi}{\langle\xi\rangle^5}.$$

Here, $tr\widehat{F}(x_0)$ is the trace of matrix $\widehat{F}(x_0)$. Let us denote by $b_1(x_0,\xi)$ and $b_2(x_0,\xi)$ the Fourier transforms of functions $\eta(|x|)|x|\Psi_1(x_0,x)$ and $\eta(|x|)\Psi_2(x_0,x)$ with respect to argument x:

$$b_1(x_0,\xi) := F(\eta(|x|)|x|\Psi_1(x_0,x)), \quad x_0 \in V,$$
$$b_2(x_0,\xi) := F(\eta(|x|)\Psi_2(x_0,x)), \quad x_0 \in V$$

Then, (3.22) can be rewritten as

$$\begin{aligned} A_V u &= \int \frac{a_0(x_0)}{\langle\xi\rangle}\widehat{u}(\xi)e^{ix_0\cdot\xi}\,d\xi - i\int \frac{B(x_0)\cdot\xi}{\langle\xi\rangle^3}\widehat{u}(\xi)e^{ix_0\cdot\xi}\,d\xi \\ &+ \int \frac{tr\widehat{F}(x_0)\langle\xi\rangle^2 - 3(\widehat{F}(x_0)\xi)\cdot\xi}{\langle\xi\rangle^5}\widehat{u}(\xi)e^{ix_0\cdot\xi}\,d\xi \\ &+ \int b_1(x_0,\xi)\widehat{u}(\xi)e^{ix_0\cdot\xi}\,d\xi + \int b_2(x_0,\xi)\widehat{u}(\xi)e^{ix_0\cdot\xi}\,d\xi, \end{aligned} \tag{3.24}$$

where $\widehat{u}(\xi)$ is the Fourier transform of function $u \in C_0^\infty(V)$ and symbols $b_1, b_2 \in C^\infty(V \times \mathrm{R}^2)$ are the Fourier transforms of finite functions. Formula (3.24) defines the PDO

$$A_V : C_0^\infty(V) \to C^\infty(V)$$

of the order -1 with the positively homogeneous (with respect to ξ) principal symbol $a_0(x_0)|\xi|^{-1}$.

The last terms in (3.22) and in (3.24) produce operators with infinitely smooth kernels, so that each of them belongs to $L^{-\infty}(V)$ [15, 40]. For the function $b_1(x_0,\xi)$, we have the estimate

$$\left|\partial_{x_0}^p \partial_\xi^q b_1(x_0,\xi)\right| \le C_{K,p,q}\langle\xi\rangle^{-2+|q|} \tag{3.25}$$

for every compact set $K \subset V$ and arbitrary multiindices p and q. Indeed,

$$\begin{aligned} \xi_j^2\left|\partial_{x_0}^p\partial_\xi^q b_1(x_0,\xi)\right| &= \frac{1}{2\pi}\left|\int \frac{\partial^{p+2}(\Psi_1(x_0,x)\eta(|x|)|x|)}{\partial x_0^p \partial x_j^2} e^{-ix\cdot\xi}\xi^q\,dx\right| \\ &= \frac{|\xi|^{|q|}}{2\pi}\left|\int\left(\frac{\partial^{p+2}(\Psi_1(x_0,x)\eta(|x|))}{\partial x_0^p\partial x_j^2}|x|\right.\right. \\ &\quad + 2\frac{\partial^{p+1}(\Psi_1(x_0,x)\eta(|x|))}{\partial x_0^p\,\partial x_j}\frac{x_j}{|x|} \\ &\quad \left.\left. + \frac{\partial^p\Psi_1(x_0,x)\eta(|x|)}{\partial x_0^p}\frac{|x|^2 - x_j^2}{|x|^3}\right)e^{-ix\cdot\xi}\,dx\right| \\ &\le \tilde{C}_{K,p,q}|\xi|^{|q|} \quad (j = 1,2) \end{aligned}$$

(since the last integral converges absolutely). This estimate yields (3.25). From (3.25), it follows that $b_1 \in S^{-2}(V)$ and the fourth term in (3.24) is a PDO belonging to the class $L^{-2}(V)$ [15, 40].

Thus, for the PDO A_V, we have the representation

$$A_V u \equiv A_V^0 u + B_V u = \int a_V(x_0,\xi)\widehat{u}(\xi)e^{ix_0\cdot\xi}\,d\xi$$
$$= a_0(x_0)\int \frac{1}{\langle\xi\rangle}\widehat{u}(\xi)e^{ix_0\cdot\xi}\,d\xi + \int b_V(x_0,\xi)\widehat{u}(\xi)e^{ix_0\cdot\xi}\,d\xi, \tag{3.26}$$

where

$$a_V(x_0,\xi) = a_0(x_0)\langle\xi\rangle^{-1} + b_V(x_0,\xi),$$

the symbol $b_V \in S^{-2}(V)$ and $B_V \in L^{-2}(V)$.

Consider the operator $\tilde{A}$ defined by the formula

$$\tilde{A}u = \sum_\alpha \psi_\alpha A_\alpha \varphi_\alpha u, \tag{3.27}$$

or

$$\tilde{A}u \equiv \tilde{A}^0 u + \tilde{B}u = \sum_\alpha \psi_\alpha A_\alpha^0 \varphi_\alpha u + \sum_\alpha \psi_\alpha B_\alpha \varphi_\alpha u, \tag{3.28}$$

where $A_\alpha \equiv A_V$, $A_\alpha^0 \equiv A_V^0$, $B_\alpha \equiv B_V$ for $V = V_\alpha$, $1 = \sum_\alpha \varphi_\alpha$, and functions ψ_α are such that $supp\,\psi_\alpha \subset V_\alpha$ and $\psi_\alpha\varphi_\alpha \equiv \varphi_\alpha$. The operator $\tilde{A} : C_0^\infty(\Omega) \to C^\infty(\Omega)$.

Since the kernel of A in (3.15) has a singularity only at $x = y$, operator A differs from $\tilde{A}$ by an operator with an infinitely smooth kernel:

$$A = \tilde{A} + \tilde{K}, \quad \tilde{K} \in L^{-\infty}(\Omega). \tag{3.29}$$

The PDO with the symbol $\langle\xi\rangle^{-1}$ performs an isomorphic mapping of $H^{-1/2}(M)$ on $H^{1/2}(M)$ [35]. Since $a_0(x_0) > 0$ in every coordinate vicinity, $C_0^\infty(\Omega)$ is dense in $\tilde{H}^{-1/2}(\overline{\Omega})$, and $\tilde{H}^{-1/2}(\overline{\Omega})$ is antidual with respect to $H^{1/2}(\Omega)$, operator $\tilde{A}^0$ can be continued up to a bounded, continuously invertible operator

$$\tilde{A}^0 : \tilde{H}^{-1/2}(\overline{\Omega}) \to H^{1/2}(\Omega).$$

Then, by virtue of (3.28) and (3.29),

$$A : \tilde{H}^{-1/2}(\overline{\Omega}) \to H^{1/2}(\Omega)$$

is a bounded Fredholm operator with the zero index.

In order to analyze the properties of operator A_τ, we will consider its restriction $\widehat{A}_V$ on V. Let $e_i(x)$ $(i = 1,2)$ be the unit basis vectors of a local coordinate system on surface U. In the local coordinates, the action of $\widehat{A}_V$ on functions $u \in C_0^\infty(V)$ can be represented as

$$\widehat{A}_V u = \sum_{i,j=1}^{2} e_j(x_0)(A_V^{ij}u_i)(x_0),$$

where

$$A_V^{ij} = \int_V \frac{e^{ik\Phi(x,x_0)|x-x_0|}}{\Phi(x,x_0)|x-x_0|} E_{ij}(x,x_0)u_i(x)\sigma(x)\,dx,$$

$$E_{ij}(x,x_0) := e_i(x)\cdot e_j(x_0), \quad u_i(x) := u(x)\cdot e_i(x),$$

and $E_{ij} \in C^{\infty}(V \times V)$ and $E_{ij}(x_0, x_0) = \delta_{ij}$ (δ_{ij} is the Kronecker delta). If A_V^{ij} is a PDO on V, then $\widehat{A}_V$ is transformed to a matrix PDO and A_τ is a PDO on the manifold Ω.

The kernels of integral operators A_V^{ij} differ from the kernels of integral operator A only by factors $E_{ij}(x, x_0)$; therefore, investigation of these operators is completely analogous to the analysis of A. We will limit ourselves to the final expression for $\widehat{A}_V$ that corresponds to (3.26):

$$\widehat{A}_V u \equiv \widehat{A}_V^0 u + \widehat{B}_V u = \int \widehat{a}_V(x_0, \xi)\widehat{u}(\xi) e^{ix_0\cdot\xi}\, d\xi$$

$$= a_0(x_0) \int \frac{1}{\langle\xi\rangle} \widehat{u}(\xi) e^{ix_0\cdot\xi}\, d\xi + \int \widehat{b}_V(x_0, \xi)\widehat{u}(\xi) e^{ix_0\cdot\xi}\, d\xi, \tag{3.30}$$

where $\widehat{a}_V = a_0(x_0)\langle\xi\rangle^{-1}\widehat{I} + \widehat{b}_V(x_0, \xi)$ ($\widehat{I}$ is the unit matrix); the matrices

$$\widehat{a}_V = \{a_V^{ij}(x_0, \xi)\}_{i,j=1}^2, \quad a_0(x_0)\langle\xi\rangle^{-1}\widehat{I}, \quad \widehat{b}_V = \left\{b_V^{ij}(x_0, \xi)\right\}_{i,j=1}^2$$

are the total symbols of the PDOs $\widehat{A}_V$, $\widehat{A}_V^0$, and $\widehat{B}_V$; and

$$\widehat{b}_V \in S^{-2}(V), \quad \widehat{B}_V \in L^{-2}(V).$$

Here, $\widehat{u}(\xi)$ is the Fourier transform of the vector-function $u = (u^1, u^2)$, $u \in C_0^{\infty}(V)$. Note that the matrix symbol of operator $\widehat{A}_V^0$ has a diagonal structure.

We define the operator $\widehat{A}$ by pasting together with the help of the formula

$$\widehat{A}u = \sum_{\alpha} \psi_\alpha \widehat{A}_\alpha \varphi_\alpha u, \tag{3.31}$$

or, in a more detailed form,

$$\widehat{A}u \equiv \widehat{A}^0 u + \widehat{B}u = \sum_{\alpha} \psi_\alpha \widehat{A}_\alpha^0 \varphi_\alpha u + \sum_{\alpha} \psi_\alpha \widehat{B}_\alpha \varphi_\alpha u, \tag{3.32}$$

where $\widehat{A}_\alpha = \widehat{A}_V$, $\widehat{A}_\alpha^0 = \widehat{A}_V^0$, $\widehat{B}_\alpha = \widehat{B}_V$, and $V = V_\alpha$.

Operator A_τ differs from $\widehat{A}$ by an operator with an infinitely smooth kernel:

$$A_\tau = \widehat{A} + \widehat{K}, \quad \widehat{K} \in L^{-\infty}(\Omega). \tag{3.33}$$

Repeating the analysis performed for operator A and using the diagonal form of the matrix symbols of operators $\widehat{A}_V$, we obtain that the operator

$$\widehat{A}^0 : \tilde{H}^{-1/2}(\overline{\Omega}) \to H^{1/2}(\Omega)$$

is continuously invertible. Then, by virtue of (3.32) and (3.33),

$$A_\tau : \tilde{H}^{-1/2}(\overline{\Omega}) \to H^{1/2}(\Omega)$$

is a bounded Fredholm operator with the zero index.

Now, let us consider the operator

$$Lu := grad_\tau\, A\,(div\, u) + k^2\, A_\tau u, \tag{3.34}$$

defined by the left-hand side of formula (3.14). Since L will be considered as an operator acting from W to W' and $C_0^{\infty}(\Omega)$ is dense in W, it is sufficient to define L on $C_0^{\infty}(\Omega)$ and to prove that this operator is bounded on this pair of spaces. Then L can be continued on entire W as a continuous operator.

The quadratic form of operator L is

$$(Lu, u) \equiv \int_\Omega Lu \cdot \overline{u}\, ds = \int_\Omega grad_\tau A(div\, u) \cdot \overline{u}\, ds + k^2 \int_\Omega A_\tau u \cdot \overline{u}\, ds. \tag{3.35}$$

We set $h(x) = A(div\, u)$, $x \in \Omega$, so that $H(X)$ is a continuation of $h(x)$ to the vicinity G_t: $H(\mathrm{x}) = h(x)$ and $H_1(\mathrm{x}) = \nu(x) \times h(x)$ for all $\mathrm{x} = x + \nu(x)t$, $|t| < \delta$. Then, $h(x) = H_1(x) \times \nu(x)$, $x \in \Omega$. Applying vector identities and the Stokes theorem, we obtain, repeating the derivation of modified representation (3.8) for the field E, the chain of equalities

$$\begin{aligned}
\int_\Omega grad_\tau A(div\, u) \cdot \overline{u}\, ds &= \int_\Omega grad_\tau H \cdot \overline{u}\, ds = \int_\Omega Grad\, H \cdot \overline{U}\, ds \\
&= \int_\Omega Div\,(H\overline{U})\, ds - \int_\Omega H\, Div\, \overline{U}\, ds \\
&= \int_\Omega Div\,(H\overline{U}_1 \times \nu)\, ds - \int_\Omega h\, div\, \overline{u}\, ds \\
&= \int_\Omega \nu \cdot Rot\,(H\overline{U}_1)\, ds - \int_\Omega h\, div\, \overline{u}\, ds \\
&= \oint_\Gamma H\overline{U}_1 \cdot t\, dl - \int_\Omega h\, div\, \overline{u}\, ds = -\int_\Omega h\, div\, \overline{u}\, ds \\
&= -\int_\Omega A\,(div\, u)(div\, \overline{u})\, ds = -(A\,(div\, u), div\, u).
\end{aligned}$$

In the above formulas, we used the notations of Section 3.1.

We have proved that

$$(Lu, u) = -(A\,(div\, u), div\, u) + k^2\,(A_\tau u, u). \tag{3.36}$$

According to the definition of space W, one can prove that form (3.36) is bounded in W if quadratic forms (Au, u) and $(A_\tau u, u)$ are bounded on $\tilde{H}^{-1/2}(\overline{\Omega})$. On the other hand, if to take into account (3.29) and (3.33), it is sufficient to show that forms $(\tilde{A}u, u)$ and $(\hat{A}u, u)$ are bounded. We limit ourselves to the proof of this fact for $(\tilde{A}u, u)$, because for $(\hat{A}u, u)$, the proof is similar.

For the form $(\tilde{A}u, u)$, we have

$$(\tilde{A}u, u) = \sum_\alpha (\psi_\alpha A_\alpha \varphi_\alpha u, u) = \sum_\alpha (A_\alpha u_\alpha, v_\alpha),$$

where $v_\alpha := \psi_\alpha u$, $supp\, v_\alpha \subset V_\alpha$, and $supp\, u_\alpha \subset V_\alpha$. But the form $(A_\alpha u_\alpha, v_\alpha)$ is bounded, because, in local coordinates, A_α is an elliptic PDO of the order -1 and class $L^{-1}(V_\alpha)$, bounded from $\tilde{H}^{-1/2}(\overline{V}_\alpha)$ to $H^{1/2}(V_\alpha)$. Consequently, the form $(A_\alpha u_\alpha, v_\alpha)$ is also bounded on $\tilde{H}^{-1/2}(\overline{V}_\alpha)$.

The quadratic form of operator L on W is bounded; therefore, sesquilinear form (Lu, v) is also bounded on W and one can consider L as a bounded operator $L : W \to W'$, where W' is the space antidual with respect to W [26].

Now, we can analyze (3.14) as a vector pseudodifferential equation

$$Lu = f, \quad u \in W, \quad f \in C^\infty(\overline{\Omega}) \subset W'. \tag{3.37}$$

The equality in (3.37) is understood in the sense of distributions.

Definition 3 *An element $u \in W$ is called a generalized solution to equation (3.37) or (3.14), if for every $v \in C_0^\infty(\Omega)$, the variational identity*

$$-(A(div\, u), div\, v) + k^2 (A_\tau u, v) = (f, v) \tag{3.38}$$

is valid.

3.4 Fredholm property and solvability of the vector pseudodifferential equation

In order to prove that $L : W \to W'$ is a Fredholm operator with the zero index, it is sufficient to represent L as a sum of continuously invertible and compact operators.

We rewrite (3.34) as

$$Lu = L_1 u + k^2 L_2 u, \tag{3.39}$$

where

$$L_1 u := grad_\tau A(div\, u), \quad L_2 u := A_\tau u,$$

and consider the action of operators L_1 and L_2 on subspaces W_1 and W_2. According to Statements 17 and 21, we have the matrix expansion

$$L_1 = \begin{pmatrix} 0 & 0 \\ 0 & \widehat{L}_1 \end{pmatrix} : \begin{pmatrix} W_1 \\ W_2 \end{pmatrix} \to \begin{pmatrix} W^1 \\ W^2 \end{pmatrix}, \tag{3.40}$$

where $\widehat{L}_1 : W_2 \to W^2$ is a bounded operator. For the operator L_2, we obtain

$$L_2 = \begin{pmatrix} \widehat{L}_{11} & \widehat{L}_{12} \\ \widehat{L}_{21} & \widehat{L}_{22} \end{pmatrix} : \begin{pmatrix} W_1 \\ W_2 \end{pmatrix} \to \begin{pmatrix} W^1 \\ W^2 \end{pmatrix}, \tag{3.41}$$

where $\widehat{L}_{ij} : W_i \to W^j$ are bounded operators. The norms on W_i and W^j are induced by the norms on W and W'.

Lemma 7 *The operator $L_0 : W_i \to W^j$ is compact if there exists an $s \in \mathrm{R}$ such that*

$$\|L_0 u\|_{H^s(\Omega)} \le C \, \|u\|_{\tilde{H}^{i-3/2}(\overline{\Omega})}, \quad s > 3/2 - j, \quad i, j = 1, 2. \tag{3.42}$$

□ By virtue of Statement 19,

$$\|\cdot\|_W \sim \|\cdot\|_{H^{i-3/2}(\overline{\Omega})} \quad \text{on} \quad W^i,$$

and, therefore,

$$\|\cdot\|_{W'} \sim \|\cdot\|_{H^{3/2-j}(\Omega)} \quad \text{on} \quad W^j.$$

But the embedding $H^s(\Omega) \subset H^{3/2-j}(\Omega)$ is compact when $s > 3/2 - j$ [58], which, together with estimate (3.42), yields the compactness of operator L_0. □

Applying formulas (3.28) and (3.29) to operator $\widehat{L}_1$, we obtain the representation

$$\begin{aligned}\widehat{L}_1 u &= grad\, \tilde{A}^0 (div\, u) + grad\, (\tilde{B} + \tilde{K})\, (div\, u) \\ &= -L_1^{(1)} u + L_1^{(2)} u. \end{aligned} \tag{3.43}$$

Since $\tilde{B}$ is a PDO of the order -2 and $\tilde{K}$ is an operator with an infinitely smooth kernel, the conditions of Lemma 1 with $i = j = 2$ and $s = 1/2$ hold for $L_1^{(2)}$, and operator $L_1^{(2)}$ is compact. The quadratic form of operator $L_1^{(1)}$ is coercive on W_2, that is

$$\begin{aligned}(L_1^{(1)} u, u) &= (\tilde{A}^0 (div\, u), div\, u) \\ &\geq C\, \|div\, u\|^2_{-1/2} \geq C_1\, \|u\|^2_W\,, \quad u \in W_2.\end{aligned}$$

Therefore, according to Ref. [26], $L_1^{(1)}$ is continuously invertible, so that $\widehat{L}_1$ is a Fredholm operator with $ind\, \widehat{L}_1 = 0$.

Since L_2 is a PDO of the order -1, operators $\widehat{L}_{ji}$, $ij \neq 1$ (or $\widehat{L}_{12}$, $\widehat{L}_{21}$, and $\widehat{L}_{22}$) satisfy the conditions of Lemma 7 with $s = i - 1/2$ and these operators are also compact.

Consider now operator $\widehat{L}_{11}$. Using the notations of Section 3.2, we have

$$\widehat{L}_{11} = L_{11}^{(1)} + L_{11}^{(2)}, \tag{3.44}$$

where

$$\begin{aligned}L_{11}^{(1)} &:= p(1 - \tilde{P}) q \widehat{A}^0 (1 - P), \\ L_{11}^{(2)} &:= p(1 - \tilde{P}) q (\widehat{B} + \widehat{K})(1 - P).\end{aligned}$$

Operator $\widehat{B} + \widehat{K}$ is a PDO of the order -2; therefore, $L_{11}^{(2)}$ satisfies the conditions of Lemma 7 with $i = j = 1$ and $s = 3/2$, so that $L_{11}^{(2)}$ is also compact. The quadratic form of operator $L_{11}^{(1)}$ is coercive on W_1,

$$(L_{11}^{(1)} u, u) = (\widehat{A}^0 u, u) \geq C\, \|u\|^2_{-1/2} = C\, \|u\|^2_W\,, \quad u \in W_1,$$

and hence $L_{11}^{(1)}$ is continuously invertible [26]. Thus, we have proved that $\widehat{L}_{11}$ is a Fredholm operator, and $ind\, \widehat{L}_{11} = 0$.

Combining the above results and taking into account formulas (3.39)–(3.44), we obtain the matrix representation for L in the form

$$L = L^1 + L^2 \equiv \begin{pmatrix} k^2 L_{11}^{(1)} & 0 \\ 0 & -L_1^{(1)} \end{pmatrix} + \begin{pmatrix} k^2 L_{11}^{(2)} & k^2 \widehat{L}_{12} \\ k^2 \widehat{L}_{21} & L_1^{(2)} + k^2 \widehat{L}_{22} \end{pmatrix}, \tag{3.45}$$

where operator L^2 is compact and operator L^1 is continuously invertible for $k \neq 0$, because operators $L_{11}^{(1)}$ and $L_1^{(1)}$ are continuously invertible. Thus, we have proved the following

Theorem 6 *The operator $L = L(k) : W \to W'$ is Fredholm for $k \neq 0$ and $ind\, L(k) = 0$.*

Remark 3 *At $k = 0$, $L(0)$ is not a Fredholm operator because $W_1 \subset ker\, L^1$ and $dim\, W_1 = \infty$.*

Remark 4 *We have already verified that $L^{(1)}_{11}$ and $L^{(1)}_{1}$ are uniformly positive operators. The matrix expansion of operator L^1 in (3.45) shows that, for $\Re k \neq 0$, the principal part of L (operator L^1) is not a positively defined or a negatively defined operator. However, L^1 is coercive for $\Im k \neq 0$:*

$$|(L^1 u, u)| \geq C\|u\|^2_W, \quad u \in W.$$

Obviously, for $\Im k = 0$ and $k \neq 0$, operator L^1 is not coercive.

Using Theorems 5 and 6, one can prove that for $\Im k \geq 0$ and $k \neq 0$, a stronger statement holds concerning the unique solvability of equation (3.14) or (3.37). In order to do this, we must verify the smoothness of solutions to equation (3.37) in Ω when $f \in C^\infty(\overline{\Omega})$.

Assume that $f = f^1 + f^2$, $f^1 \in W^1$, $f^2 \in W^2$, and $f \in C^\infty(\overline{\Omega})$. Taking into account matrix representation (3.45), we rewrite equation (3.37) as a system of two equations ($k \neq 0$):

$$L^{(1)}_{11} u_1 = -L^{(2)}_{11} u_1 - \widehat{L}_{12} u_2 + k^{-2} f^1, \tag{3.46}$$

$$L^{(1)}_{1} u_2 - L^{(2)}_{1} u_2 + k^2 \widehat{L}_{21} u_1 + k^2 \widehat{L}_{22} u_2 - f^2, \tag{3.47}$$

where $u_1 \in W_1$, $u_2 \in W_2$, and $u = u_1 + u_2$. Operators $\widehat{L}_{ij} : W_i \to W^j$ $(i, j \neq 1)$ are PDOs of the order -1 and class $L^{-1}(\Omega)$, $L^{(2)}_{11} : W_1 \to W^1$ is a PDO of the order -2 and class $L^{-2}(\Omega)$, and $L^{(2)}_{1} : W_2 \to W^2$ is a PDO of the order 0 and class $L^0(\Omega)$. Let us denote the right-hand sides of (3.46) and (3.47) by g^1_* and g^2_*, respectively. Then,

$$L^{(1)}_{11} u_1 = g^1_*, \quad g^1_* \in H^{3/2}(\Omega), \tag{3.48}$$

$$L^{(1)}_{1} u_2 = g^2_*, \quad g^2_* \in H^{1/2}(\Omega). \tag{3.49}$$

Here, $L^{(1)}_{11} : W_1 \to W^1$ is a classical elliptic PDO of the order 1 and class $L^{-1}_{\mathrm{cl}}(\Omega)$ with the principal symbol $\widehat{a}_0(x,\xi) = a_0(x)|\xi|^{-1}\widehat{I}$. The principal symbol $\widehat{a}_0 = \widehat{a}_0(x,\xi)$ of the operator specifies, for every nonzero element $(x,\xi) \in T^*M$ of the cotangential bundle T^*M, the mapping of layers

$$\widehat{a}_0(x,\xi) : W_x \to W'_x,$$

so that, finally, we obtain the mapping of bundles $\widehat{a}_0 : \pi_0^* W \to \pi_0^* W'$, where $\pi_0 : T^*M \setminus O \to M$ is the canonical projection of the cotangential bundle without the zero cross section on the base M, and $\pi_0^* W$ and $\pi_0^* W'$ are the induced bundles with the layers W_x and W'_x over every point $(x,\xi) \in T^*M \setminus O$.

Consider the operator $L^{(1)}_{1} : W_2 \to W^2$, $L^{(1)}_{1} u := grad\, \tilde{A}^0 (div\, u)$. This operator is a composition of two differential operators *div* and *grad* and a scalar PDO $\tilde{A}^0$ with the principal symbol $a_0(x)|\xi|^{-1}$, which is a correctly defined function on cotangential bundle T^*M. Using local coordinates to represent the differential operators in the vicinities $U = U_\alpha$ and $V = V_\alpha$, respectively, we obtain

$$div\, u = \int e^{ix\cdot\xi} G^{-1} \left(i\xi_1 \widehat{u}^1 \sqrt{G} + i\xi_2 \widehat{u}^2 \sqrt{G} + \widehat{u}^1 \frac{\partial \sqrt{G}}{\partial x_1} + \widehat{u}^2 \frac{\partial \sqrt{G}}{\partial x_2} \right) d\xi,$$

$$u = (u^1, u^2)^T,$$

$$grad\, g = \int e^{ix\cdot\xi} G^{-1/2} (i\xi_1 \widehat{g}, i\xi_2 \widehat{g})^T\, d\xi.$$

Note that *div* and *grad* are (differential) eigenoperators and $\tilde{A}^0$ is a classical PDO; therefore, their composition can be defined also as a classical PDO. According to the theorem about the composition of classical PDOs [15, 40], the principal symbol $\widehat{\sigma}_0$ of operator $L_1^{(1)}$ can be determined as a product of principal symbols of these operators,

$$\widehat{\sigma}_0(x,\xi) = -\frac{1}{a_0(x)|\xi|}\begin{pmatrix} \xi_1^2 & \xi_1\xi_2 \\ \xi_2\xi_1 & \xi_2^2 \end{pmatrix}. \tag{3.50}$$

Formally, symbol (3.50) degenerates for all $\xi \in \mathrm{R}^2$, since the matrix determinant in (3.50) identically equals zero.

However, operator $L_1^{(1)}$ acts on subspace W_2 with elements u satisfying the condition $rot_\nu u = 0$. In local coordinates, for $u \in C_0^\infty(V)$, this condition is equivalent to

$$\frac{\partial u^2}{\partial x_1} - \frac{\partial u^1}{\partial x_2} + u^2 b_1 - u^1 b_2 = 0, \quad x \in V, \tag{3.51}$$

where $b_j = \psi\left(G^{-1/2}\partial G^{1/2}/\partial x_j\right)$ and the functions $\psi = \psi_\alpha$ are introduced in (3.27). Calculating the Fourier transform of the left-hand side of equality (3.51), we obtain

$$i\xi_1\widehat{u}^2 - i\xi_2\widehat{u}^1 + \widehat{u^2 b_1} - \widehat{u^1 b_2} = 0, \quad \xi \in \mathrm{R}^2. \tag{3.52}$$

Then, for u that satisfy condition (3.52), we have

$$\widehat{\sigma}_0(x,\xi)\widehat{u}(\xi) = -\frac{|\xi|}{a_0(x)}\widehat{u}(\xi) + \frac{1}{a_0(x)|\xi|}i\xi'\widehat{b \cdot u}(\xi), \tag{3.53}$$

where $b := (b_2, -b_1) \in C_0^\infty(V)$ and $\xi' = (\xi_2, -\xi_1)$. Combining these formulas and taking into account that the operator of multiplication by a smooth vector b preserves the class of a PDO, we obtain that $L_1^{(1)}$ acts on subspaces $W_2 \to W^2$ as a classical elliptic PDO of the order 1 and class $L^1_{\mathrm{cl}}(\Omega)$ and has the principal symbol $-a_0^{-1}(x)|\xi|\widehat{I}$. Thus,

$$\begin{aligned} L_{11}^{(1)} &= a_0\Delta^{-1/2} &&: W_1 \to W^1, \\ L_1^{(1)} &= -a_0^{-1}\Delta^{1/2} &&: W_2 \to W^2, \end{aligned} \tag{3.54}$$

where $\Delta^{\pm 1/2}$ are classical PDOs with the principal symbols $|\xi|^{\mp 1}$.

Statement 22 *If $u \in W$ is a solution to equation (3.14) with a smooth right-hand side $f \in C^\infty(\overline{\Omega})$, then $u \in C^\infty(\Omega)$.*

□ The proof is performed in a standard manner, by a step-by-step smoothing of solutions u_1 and u_2 of equations 3.46) and (3.47). The possibility of smoothing follows from the existence of the parametrix for elliptic PDOs $L_{11}^{(1)} \in L^{-1}_{\mathrm{cl}}(\Omega)$ and $L_1^{(1)} \in L^1_{\mathrm{cl}}(\Omega)$. Indeed, equations (3.48) and (3.49) yield [15] $u_1 \in H^{1/2}_{loc}(\Omega)$ and $u_2 \in H^{3/2}_{loc}(\Omega)$. Considering equations (3.46) and (3.47) and taking into account that, according to Ref. [15], PDOs of the class $L^m(\Omega)$ act continuously in the spaces

$$H^s_{comp}(\Omega) \to H^{s-m}_{loc}(\Omega), \quad s \in \mathrm{R},$$

we obtain equations (3.48) and (3.49) with the right-hand sides $g_*^1 \in H^{5/2}_{loc}(\Omega)$ and $g_*^2 \in H^{3/2}_{loc}(\Omega)$. Repeating this procedure, we conclude that $u_1, u_2 \in H^s_{loc}(\Omega)$ for every s, which completes the proof. □

Theorem 7 *For $\Im k \geq 0$ and $k \neq 0$, the operator $L(k) : W \to W'$ is continuously invertible.*

□ In order to prove the required property of $L(k)$, it is sufficient to show, according to Theorem 6, that $\ker L(k) = \{0\}$ for $\Im k \geq 0$ and $k \neq 0$. Let $u_0 \in \ker L(k)$. Then, as follows from Statement 22, u satisfies conditions (3.6) and (3.7). But, by virtue of Theorem 5, the equation $L(k)u_0 = 0$ has only the trivial solution that satisfies (3.6) and (3.7). Thus, $u_0 \equiv 0$ and the theorem is proved. □

Corollary 3 *For $\Im k \geq 0$ and $k \neq 0$, there exists the unique generalized solution $u \in W$ of equation (3.14) or (3.37) for arbitrary right-hand side $f \in W'$ and, in particular, for $f \in C^\infty(\overline{\Omega})$.*

Let us summarize our study of the solvability of the diffraction problem for a system Ω of arbitrary bounded screens.

Theorem 8 *Problem (2.1)–(2.6) is uniquely solvable for $\Im k \geq 0$, $k \neq 0$, and arbitrary E^0, H^0 satisfying condition (2.7).*

□ In order to prove this theorem, it is sufficient to repeat the proof of Theorem 4. □

Corollary 4 *Every solution to problem (2.1)–(2.6) can be represented, for $\Im k \geq 0$ and $k \neq 0$, as vector potential (3.3)–(3.5), where u satisfies conditions (3.6) and (3.7).*

In this chapter, we will not perform a detailed analysis of the smoothness of solution $u \in W$ in the vicinity of the edge Γ for a smooth right-hand side $f \in C^\infty(\overline{\Omega})$. Here, it is easy to reduce the consideration, by means of the straightening procedure, to the case of a planar screen examined in Chapter 2. In this section, we will limit ourselves to a rough (but the most important) result concerning singulatrity of a solution in the vicinity of the edge Γ.

Taking into account (3.54), we write equation (3.48), (3.49) in the form

$$\Delta^{-1/2} u_1 = \tilde{g}^1_*, \quad \tilde{g}^1_* \in H^{3/2}(\Omega), \tag{3.55}$$

$$\Delta^{1/2} u_2 = \tilde{g}^2_*, \quad \tilde{g}^2_* \in H^{1/2}(\Omega). \tag{3.56}$$

The smoothness of solutions to these equations (on a manifold) were considered in Refs. [87, 88]. Performing the analysis similar to the proof of Lemma 6, one can represent the solutions in the form

$$u_1 = u_1^0 + K(-1/2, 0)v^1, \tag{3.57}$$
$$u_1^0 \in \tilde{H}^{1/2-\varepsilon}(\overline{\Omega}), \quad v^1 \in H^{1/2-\varepsilon}(\Gamma),$$

$$u_2 = u_2^0 + K(1/2, 0)v^2, \tag{3.58}$$
$$u_2^0 \in \tilde{H}^{3/2-\varepsilon}(\overline{\Omega}), \quad v^2 \in H^{3/2-\varepsilon}(\Gamma),$$

where ε is an arbitrarily small positive number and operator $K(p, 0)$ that acts on $H^s(\Gamma)$, $(K(p, 0)v)(x) \in \Omega$, is defined with the help of (2.49) and the partition of unity in local coordinates (see Ref. [88]).

From (3.57) and (3.58), it follows that, in the vicinity of Γ, the leading singular term of solution u has the form

$$u \sim \rho^{-1/2}(x)\, g(t), \quad g \in H^{1/2-\varepsilon}(\Gamma), \tag{3.59}$$

where $\rho(x) = dist\,(x, \Gamma) = |x - t|$, $t \in \Gamma$ in local coordinates. The order of singularity will be obviously preserved if to pass on to global coordinates on Ω.

3.5 Principle of limiting absorption

The properties of solutions to the diffraction problem on Ω, established in the previous sections, enable us to prove a number of important statements concerning the dependence of solutions on parameter k and first of all, the possibility of a limiting transition $k \to k_0$, where k is a complex number with $\Im k > 0$ and $k_0 \neq 0$ is a real number. Such a limiting transition constitutes the principle of limiting absorption.

The principle of limiting absorption may be formulated both as a limiting transition for a solution $u = u(k)$ of equation (3.14) on Ω and as a limiting transition for the scattered field $E = E(k)$, $H = H(k)$. In every case, it is assumed that the incident field $E^0 = E^0(k)$, $H^0 = H^0(k)$ is continuous with respect to k (which is usually assumed in practical applications). It is sufficient to suppose that such a dependence is valid only in the vicinity of the point k_0 and for a k with a positive imaginary part, $k \to k_0$ and $\Im k > 0$. Below, we will give exact formulations of the conditions imposed on the incident field.

Theorem 7 and analytical dependence of the operator-valued function $L(k)$ on the parameter $k \in \mathrm{C}$ serve as a basis for the proof of the principle of limiting absorption. We say that an operator-valued function $L(k)$ is analytical (holomorphic) in a domain D if $L(k)$ is differentiable with respect to the norm at every point of D.

Statement 23 *The operator-valued function $L(k) : W \to W'$ is analytical (holomorphic) for all $k \in \mathrm{C}$.*

□ The simplest way to verify that $L(k)$ is holomorphic is to apply the criterion proved in Ref. [26, p. 195]: in order to prove that $L(k)$ is holomorphic (with respect to the norm), it is sufficient to check that $(L(k)u, v)$ is an analytical function of the variable $k \in \mathrm{C}$ for all $u, v \in C_0^\infty(\Omega)$, which is dense in W, and that $(L(k)u, v)$ is locally bounded with respect to k for every fixed pair of $u, v \in W$. Formula (3.36) yields

$$(L(k)u, v) = -\left(A\,(div\,u), div\,v\right) + k^2\left(A_\tau u, v\right),$$

where A and A_τ are given by formula (3.15). All the above conditions are obviously valid for this function. □

Statement 24 *The operator-valued function $L^{-1}(k) : W' \to W$ is analytical (holomorphic) in C_+ (for $\Im k > 0$) and continuous in $\overline{\mathrm{C}}_+ \setminus 0$ (for $\Im k \geq 0$ and $k \neq 0$).*

□ $L(k)$ is analytical in C_+ because

$$\frac{d}{dk}(L^{-1}(k)) = -L^{-1}(k)\left(\frac{d}{dk}L(k)\right)L^{-1}(k), \tag{3.60}$$

Note that, by virtue of Theorem 7 and Statement 23, the right-hand side of (3.60) is defined; therefore, the left-hand side is also a well-defined function of k. The identity

$$L^{-1}(k) = L^{-1}(k_0)\left(I + (L(k) - L(k_0))L^{-1}(k_0)\right)^{-1} \tag{3.61}$$

proves the continuity of $L(k)$ in $\overline{\mathrm{C}}_+ \setminus 0$. Note that this identity is valid for k and k_0 such that

$$\|L(k) - L(k_0)\| < \|L^{-1}(k_0)\|^{-1}. \tag{3.62}$$

□

Theorem 9 *Assume that $k_0 \neq 0$, $\Im k_0 = 0$, and $\Im k > 0$. Then, $f(k) \xrightarrow{W'} f(k_0)$ as $k \to k_0$ yields $u(k) \xrightarrow{W} u(k_0)$ as $k \to k_0$, where $u(k)$ and $u(k_0)$ are the solutions to equation (3.14) at k and k_0, respectively.*

□ Since $u(k) = L^{-1}(k)f(k)$ and $u(k_0) = L^{-1}(k_0)f(k_0)$, we can use the conditions of the theorem to obtain

$$\|u(k) - u(k_0)\|_W = \left\|L^{-1}(k)(f(k) - f(k_0)) + (L^{-1}(k) - L^{-1}(k_0))f(k_0)\right\|_W$$

$$\leq \left\|L^{-1}(k)\right\| \|f(k) - f(k_0)\|_{W'} + \left\|L^{-1}(k) - L^{-1}(k_0)\right\| \|f(k_0)\|_{W'} \to 0$$

as $k \to k_0$. □

Theorem 10 *Assume that $k_0 \neq 0$, $\Im k_0 = 0$, and $\Im k > 0$. Then, $f(k) \xrightarrow{W'} f(k_0)$ as $k \to k_0$ yields $E(k) \to E(k_0)$ and $H(k) \to H(k_0)$ in $L^2_{loc}(\mathrm{R}^3)$ as $k \to k_0$, where $E(k)$, $H(k)$ and $E(k_0)$, $H(k_0)$ are solutions to problem (2.1)–(2.6) at k and k_0, respectively.*

□ Using the results of Section 3.4 (Corollary 4), we prove that the solutions $E(k), H(k)$ and $E(k_0), H(k_0)$ to problem (2.1)–(2.6) can be represented as vector potentials (3.3)–(3.5) or (3.8), (3.4), and (3.5) with functions $u(k)$ and $u(k_0)$ that solve equation (3.14) at k and k_0 and satisfy conditions (3.6) and (3.7). Introduce the linear operator

$$Z := \begin{pmatrix} ik^{-1}(Grad\, A_1(div\,(\cdot)) + k^2 A_1(\cdot)) \\ Rot\, A_1(\cdot) \end{pmatrix} : W(\overline{\Omega}) \to L^2_{loc}(\mathrm{R}^3). \tag{3.63}$$

As has been shown in Section 3.2, this operator is continuous on this pair of spaces. By virtue of (3.8), (3.4), and (3.5) and conditions (3.6) and (3.7),

$$\begin{pmatrix} E \\ H \end{pmatrix} = Zu.$$

Therefore,

$$\|E\|_{L^2_{loc}(\mathrm{R}^3)} + \|H\|_{L^2_{loc}(\mathrm{R}^3)} \leq C\,\|u\|_W\,.$$

The operator $Z = Z(k)$ depends on k. From (3.63), it follows that $Z(k)$ is a continuous operator-valued function of k in $\mathrm{C} \setminus 0$ (with respect to the norm) if $A_1(k)$ is continuous with respect to k in C (with respect to the norm), where

$$A_1(k) : \tilde{H}^{-1/2}(\overline{\Omega}) \to H^1_{loc}(\mathrm{R}^3).$$

If $k_0 \neq 0$ is fixed, then, the kernel

$$\Phi(\mathrm{x}, y; k, k_0) = \frac{e^{ik|\mathrm{x}-y|} - e^{ik_0|\mathrm{x}-y|}}{|\mathrm{x} - y|}$$

of the integral operator

$$(A_1(k) - A_1(k_0))\,u = \frac{1}{4\pi} \int\limits_{\Omega} \Phi(\mathrm{x}, y; k, k_0) u(y)\, ds$$

can be estimated as

$$
\begin{aligned}
|\Phi(\mathrm{x}, y; k, k_0)| &\leq C|k-k_0|, \\
\left|\frac{\partial}{\partial \mathrm{x}_j}\Phi(\mathrm{x}, y; k, k_0)\right| &\leq C|k-k_0|, \\
\left|\frac{\partial}{\partial y_i}\Phi(\mathrm{x}, y; k, k_0)\right| &\leq C\,|k-k_0|\,, \\
\left|\frac{\partial^2}{\partial \mathrm{x}_j \partial y_i}\Phi(\mathrm{x}, y; k, k_0)\right| &\leq C\frac{|k-k_0|}{|\mathrm{x}-y|},
\end{aligned}
\tag{3.64}
$$

and all estimates are uniform with respect to x, y, k, and k_0 if x and y and k and k_0 belong to compact sets in R^3 and C, respectively. For an arbitrary ball B_R of the radius R and the center at the origin, we have

$$
\begin{aligned}
&\|(A_1(k)-A_1(k_0))u\|^2_{H^1(B_R)} \\
&\leq C_0 \int_{B_R} \left(\sum_{j=1}^{3} \left\| \frac{\partial}{\partial \mathrm{x}_j}\Phi(\mathrm{x},\cdot) \right\|^2_{H^{1/2}(\Omega)} + \|\Phi(\mathrm{x},\cdot)\|^2_{H^{1/2}(\Omega)} \right) \|u\|^2_{\tilde{H}^{-1/2}(\overline{\Omega})}\, d\mathrm{x} \\
&\leq C_0\, \|u\|^2_{\tilde{H}^{-1/2}(\overline{\Omega})} \int_{B_R} \left(\sum_{j=1}^{3} \left\| \frac{\partial}{\partial \mathrm{x}_j}\Phi(\mathrm{x},\cdot) \right\|^2_{H^{1}(\Omega)} + \|\Phi(\mathrm{x},\cdot)\|^2_{H^{1}(\Omega)} \right) d\mathrm{x} \\
&\leq C_1\, |k-k_0|^2\, \|u\|_{\tilde{H}^{-1/2}(\overline{\Omega})} \int_{B_R} \left[\left(\int_\Omega \frac{1}{|\mathrm{x}-y|}\, ds \right)^2 + \left(\int_\Omega ds \right)^2 \right] d\mathrm{x} \\
&= C_2\, |k-k_0|^2\, \|u\|_{\tilde{H}^{-1/2}(\overline{\Omega})}\,.
\end{aligned}
$$

Thus, $\|A_1(k)-A_1(k_0)\| \leq C_3|k-k_0|$, which means that $A_1(k)$ is continuous with respect to $k \in \mathrm{C}$. Therefore, $Z(k)$ is a continuous operator-valued function of k in $\mathrm{C} \setminus 0$ (with respect to the norm) and norms $\|Z(k)\|$ are locally bounded. Then,

$$
\begin{aligned}
&\frac{1}{\sqrt{2}}\left(\|E(k)-E(k_0)\|_{L^2_{loc}(\mathrm{R}^3)} + \|H(k)-H(k_0)\|_{L^2_{loc}(\mathrm{R}^3)} \right) \\
&\leq \|Z(k)(u(k)-u(k_0))\|_{L^2_{loc}(\mathrm{R}^3)} + \|(Z(k)-Z(k_0))u(k_0)\|_{L^2_{loc}(\mathrm{R}^3)} \\
&\leq \|Z(k)\|\, \|u(k)-u(k_0)\|_W + \|Z(k)-Z(k_0)\|\, \|u(k_0)\|_W \to 0
\end{aligned}
$$

as $k \to k_0$, since the first and the second terms tend to zero according to Theorem 9 and the above estimates, respectively, □

Remark 5 *The condition $f(k) \xrightarrow{W'} f(k_0)$ is fulfilled if functions $f(\mathrm{x};k)$ and $\partial f(\mathrm{x};k)/\partial \mathrm{x}_j$ are continuous with respect to k in a semivicinity of k_0, or, more exactly, for $|k-k_0|<\delta$, $\Im k \geq 0$, because the norm in $C^1(\overline{\Omega})$ is stronger than the norm in $W'(\Omega)$. By virtue of the embedding theorem [57] and Statement 20, we have*

$$\|f\|_{W'} \leq C_1\, \|f\|_{1/2} \leq C_2\, \|f\|_{c^1}\,.$$

□

Thus, Theorems 9 and 10 establish the principle of limiting absorption for the diffraction problem on Ω. From the physical viewpoint, this principle means continuous dependence of the solution to this problem on conductivity σ (where $\sigma \geq 0$), Therefore, in order to construct the solution in an "absorptionless" medium ($\sigma = 0$), a small parameter $\sigma > 0$ (absorption) is sometimes introduced, and then a limiting transition $\sigma \to 0$ is performed or an approximate solution is derived.

One more important problem that arises in various applications concerns scattering frequencies of the diffraction problem, that is, complex numbers k, at which there exist nontrivial solutions to homogeneous equation (3.14). Determination of scattering frequencies, their interpretation, and application for solving time-dependent problems are considered in Refs. [45, 89]. Scattering frequencies coincide with characteristic numbers of the operator-valued function $L(k)$, namely, with $k \in \mathrm{C}$ at which there exist nontrivial solutions u of the equation $L(k)u = 0$.

Statement 25 *Characteristic numbers of the operator-valued function $L(k)$ form a discrete set in C_- with possible accumulation points at the origin and (or) at infinity, and have finite algebraic multiplicities.*

□ This statement is a direct corollary of the known theorem concerning the properties of the spectrum of a Fredholm operator-valued function [7]. Indeed, according to Theorem 6 and Statement 23, $L(k)$ is a holomorphic and Fredholm operator-valued function with $ind\, L(k) = 0$, $k \in C \setminus 0$, and $L(k)$ is continuously invertible in $\overline{\mathrm{C}}_+ \setminus 0$. □

A discrete set in C_- means that every compactum in C_- may contain a finite (in particular, zero) number of characteristic numbers and they all have the negative imaginary part. Hence, the set of characteristic numbers is either finite (and maybe empty), or countable. The multiplicity of a characteristic number of an operator-valued function is defined in Ref. [7].

Chapter 4

The problems of diffraction in domains connected through a hole in a screen

In this chapter, we consider three electromagnetic problems describing volumes connected through a hole: diffraction by a hole in a planar perfectly conducting screen, by a partially shielded layer, and by an aperture in a semi-infinite waveguide. Electrodynamic parameters in different domains may be different, and all these problems are three-dimensional and vector and cannot be reduced to scalar problems.

For a long time, these problems have been of considerable interest and have become standard. They are close to problems of diffraction by bounded screens; for example, the first problem of diffraction by a hole in a planar screen is called dual with respect to the problem of diffraction on a planar screen [2]). The main difference is that in order to construct representations of solutions, we use not the (three-dimensional) fundamental solution for the whole space, but the Green's functions of the corresponding domains. In addition to this, the proof of the uniqueness theorems is different. Nevertheless, all problems are reduced to vector pseudodifferential equations of a single family, similar to that considered in Chapter 2.

We consider three electromagnetic problems of diffraction in a half-space connected through a hole with a half-space, a layer, and a semi-infinite rectangular cylinder. These unbounded domains were chosen because, first, they describe three basic types of the field behavior at infinity and, second, for them, explicit representations of the Green's functions are available.

In Section 4.1, we investigate the Green's functions of three canonical domains: a half-space, a layer, and a semi-infinite rectangular cylinder. We separate singularities of the Green's functions and study their dependence on the parameter k. It should be noted that we perform complete separation of singularities in closed domains (including the boundary) and that these singularities are different for different domains. We propose a practical algorithm for separating singularities of the traces of the Green's functions.

In Section 4.2, we analyze the basic integrodifferential equation on Ω, to which the diffraction problems are reduced. We prove the Fredholm property with the zero index for the corresponding operator in spaces $W \to W'$. The smoothness of solutions is established for smooth right-hand sides, as well as the dependence of solutions on parameters k_1 and k_2, which are used in the proof of the principle of limiting absorption in the diffraction problems.

In Section 4.3, we consider the problem of diffraction by a hole in a planar screen. The solution is represented as a vector potential. The uniqueness theorem is proved, and then, the unique solvability of the integrodifferential equation on Ω, the diffraction problem as a whole, and the principle of limiting absorption.

This procedure is applied in Section 4.4 to the problem of diffraction on a partially shielded layer with the Sveshnikov–Werner conditions at infinity. The uniqueness theorem for the diffraction problem is proved. The unique solvability of the integrodifferential equation on Ω and of the diffraction problem as a whole are established, together with the principle of limiting absorption, for the values of parameters, at which the Green's function of the layer exist.

In Section 4.5, we consider the problem of diffraction on an aperture in a semi-infinite waveguide. Here, the main difference from the problem considered in Section 4.4 is that we apply the matrix Green's functions and different conditions at infinity in the waveguide domain. Other steps of the proofs of the solvability and uniqueness theorems repeat similar considerations developed in previous sections.

4.1 Green's functions for canonical domains

Consider the Green's functions of the Helmholtz equations for some simplest unbounded domains in R^3: a half-space, a layer, and a semi-infinite rectangular cylinder. We will call them canonical domains since they characterize three types of the field behavior at infinity. Below, we will specify boundary conditions on the boundaries of the domains. The Green's function is defined in Ref. [4]. We will assume that the wavenumber k is such that $\Im k \geq 0$ and $k \neq 0$.

1. Halfspace R^3_+.

The Green's function $G^R = G^R(\mathrm{x}, y)$, $\mathrm{x}, y \in \mathrm{R}^3_+$, is defined as a solution to the boundary value problem (where $y \in \mathrm{R}^3_+$ is fixed)

$$\Delta G^R + k^2 G^R = -\delta(\mathrm{x} - y), \quad \mathrm{x} \in \mathrm{R}^3_+, \tag{4.1}$$

$$\left.\frac{\partial G^R}{\partial \mathrm{x}_3}\right|_{\mathrm{x}_3=0} = 0, \tag{4.2}$$

with the conditions at infinity

$$G^R, Grad\, G^R = o(r^{-1}) \quad \text{for} \quad \Im k > 0, \tag{4.3}$$

$$\frac{\partial G^R}{\partial r} - ikG^R = o(r^{-1}), \quad G^R = O(r^{-1}) \quad \text{for} \quad \Im k = 0,\ k \neq 0. \tag{4.4}$$

Here, $r = |\mathrm{x}| \to \infty$ uniformly along all directions $\mathrm{x}/|\mathrm{x}|$, $\mathrm{x} \in \mathrm{R}^3_+$, and uniformly with respect to $y \in Q$, where Q is an arbitrary bounded set in R^3_+. G^R has the form [4]

$$G^R(\mathrm{x}, y) = \frac{1}{4\pi}\frac{e^{ik|\mathrm{x}-y|}}{|\mathrm{x} - y|} + \frac{1}{4\pi}\frac{e^{ik|\mathrm{x}-y^*|}}{|\mathrm{x} - y^*|}, \tag{4.5}$$

where $y^* = (y_1, y_2, -y_3)$. Obviously, G^R is analytical for all $k \in \mathrm{C}$. Note that the Green's function of the lower halfspace R^3_- is also expressed by formula 4.5, but $\mathrm{x}, y \in \mathrm{R}^3_-$.

2. Layer $U = \{x : -1 < x_3 < 0\}$.

The Green's function $G^U = G^U(x, y)$, $x, y \in U$, is defined as a solution to the boundary value problem (where $y \in U$ is fixed)

$$\Delta G^U + k^2 G^U = -\delta(x - y), \quad x \in U, \tag{4.6}$$

$$\left.\frac{\partial G^U}{\partial x_3}\right|_{x_3=0} = \left.\frac{\partial G^U}{\partial x_3}\right|_{x_3=-1} = 0, \tag{4.7}$$

with the conditions at infinity [85]

$$\frac{\partial g_n}{\partial \rho} - ik_n g_n = o(\rho^{-1/2}), \quad g_n = O(\rho^{-1/2}) \tag{4.8}$$

for g_n, $n \geq 0$, where

$$k_n^2 = k^2 - \pi^2 n^2, \quad \Im k_n \geq 0, \tag{4.9}$$

$k_n > 0$ for $k > \pi n$, $k_n < 0$ for $k < -\pi n$,

$$g_n := 2 \int_{-1}^{0} G^U(x, y) \cos \pi n x_3 \, dx_3, \tag{4.10}$$

and $\rho := (x_1^2 + x_2^2)^{1/2} \to \infty$ uniformly along all directions x'/ρ, where $x' = (x_1, x_2)$, and uniformly with respect to y belonging to an arbitrary bounded subset in $\overline{U}$. (4.8) and (4.9) constitute the Sommerfeld conditions for a bounded two-dimensional domain, The Fourier coefficients (4.10) satisfy, for a certain ρ_0 in the domain $\rho > \rho_0$, the two-dimensional Helmholtz equation with the parameter k_n^2. Hence, condition (4.8) yields [23], that g_n and $\partial g_n/\partial \rho$ decay exponentially as $\rho \to \infty$ if $k^2 < \pi^2 n^2$ ($\Im k_n > 0$).

The Green's function G^U can be represented as [85]

$$G^U(x, y) = \frac{i}{2} \sum_{n=0}^{\infty} \frac{1}{1 + \delta_{0n}} \cos \pi n x_3 \cos \pi n y_3 \, H_0^{(1)}(k_n |x' - y'|) \tag{4.11}$$

for $x' \neq y'$, where $x' = (x_1, x_2)$ and $y' := (y_1, y_2)$, or

$$G^U(x, y) = \frac{1}{4\pi} \sum_{j=-\infty}^{\infty} \left(\frac{\exp(ik|x - y - 2je_3|)}{|x - y - 2je_3|} + \frac{\exp(ik|x - y^* + 2je_3)}{|x - y^* + 2je_3|} \right), \tag{4.12}$$

where $e_3 = (0, 0, 1)$. Here, $H_0^{(1)}(z)$ is the zero-order Hankel function of the first kind and δ_{ij} is the Kronecker delta. Both representations are valid for $\Im k \geq 0$ and $k \neq \pi n$, $n \in \mathbb{Z}$. The equivalence of (4.11) and (4.12) is proved in Refs. [79, 85].

Below, we will use the properties of the trace of function G^U at $x_3 = 0$ and $y_3 = 0$. Let us apply (4.12) to separate the singularity of $G^U(x', y')$ for $|x' - y'| \to 0$:

$$\begin{aligned}
\left.G^U\right|_{x_3=y_3=0} &= G^U(x', y') \\
&= \frac{1}{2\pi} \sum_{j=-\infty}^{\infty} \frac{\exp(ik(|x' - y'|^2 + 4j^2)^{1/2})}{(|x' - y'|^2 + 4j^2)^{1/2}} \\
&= \frac{1}{2\pi} \frac{e^{ik|x'-y'|}}{|x' - y'|} + \frac{1}{\pi} \sum_{j=1}^{\infty} \frac{e^{2ikj}}{2j} \\
&\quad + \frac{1}{\pi} \sum_{j=1}^{\infty} \left(\frac{\exp(ik(|x' - y'|^2 + 4j^2)^{1/2})}{(|x' - y'|^2 + 4j^2)^{1/2}} - \frac{e^{2ikj}}{2j} \right) \\
&= \frac{1}{2\pi} \frac{e^{ik|x'-y'|}}{|x' - y'|} - \frac{1}{2\pi} \ln(1 - e^{2ik}) + B(x', y'),
\end{aligned} \tag{4.13}$$

where
$$B(x',y') := \frac{1}{\pi}\sum_{j=1}^{\infty}\left(\frac{\exp(ik(|x'-y'|^2+4j^2)}{(|x'-y'|^2+4j^2)^{1/2}} - \frac{e^{2ikj}}{2j}\right).$$

Here, $\ln z$ denotes analytical continuation of the real function $\ln t$, $t > 0$, to the set $C\setminus(-i\infty, 0]$. For the coefficients $b_j(x',y')$ of the series $B(x',y')$ and their derivatives with respect to x_i and y_i of arbitrary order α, we have the estimates [85]

$$|D^\alpha b_j(x',y')| \le C_\alpha j^{-2}, \quad x, y \in \overline{\Omega} \tag{4.14}$$

that are valid on every compact set $\overline{\Omega} \subset R^2$. Consequently, we have proved that $B \in C^\infty(\overline{\Omega}\times\overline{\Omega})$.

Using representation (4.12), one can easily verify that the function $G^U = G^U(x, y; k)$ is analytical with respect to k in C_+ ($\Im k > 0$). Denote by $\Lambda(U)$ the set of k, at which the Green's function G^U is not defined:

$$\Lambda(U) = \{k : k = \pi n,\ n \in Z\}.$$

As has been shown in Ref. [85], $G^U(x, y; k)$ is continuously differentiable with respect to x_i and y_i arbitrary number of times in $\overline{U}\times\overline{U}\setminus\{(x,x) : x \in \partial U\}$ and is a continuous function of k in $\overline{C}_+\setminus\Lambda(U)$.

3. Semi-infinite rectangular cylinder

$$T^+ = \{x : |x_1| < a/2, ||x_2| < b/2, x_3 > 0\}.$$

We will consider two Green's functions $G^T_{a,b} = G^T_{a,b}(x, y)$, $x, y \in T^+$, with different boundary conditions. They are defined as solutions to the boundary value problems formulated below. G^T_a satisfies ($y \in T^+$ is fixed) the Helmholtz equation

$$\Delta G^T_a + k^2 G^T_a = -\delta(x-y), \quad x \in T^+, \tag{4.15}$$

the boundary conditions

$$\left.\frac{\partial G^T_a}{\partial x_3}\right|_{x_3=0} = \left.G^T_a\right|_{x_1=\pm\frac{a}{2}} = \left.\frac{\partial G^T_a}{\partial x_2}\right|_{x_2=\pm\frac{b}{2}} = 0, \tag{4.16}$$

and the Sveshnikov partial radiation conditions at infinity, according to which the Green's function can be represented, for $|x_3| \ge C_0$ and a certain $C_0 > 0$, in the form of the series

$$G^T_a = \sum_p T_p e^{i\gamma_p^{(a)}|x_3|}\tilde{\Pi}_p(x_1, x_2), \tag{4.17}$$

$$\gamma_p^{(a)} = \sqrt{k^2 - \lambda_p^{(a)}},\ \Im\gamma_p^{(a)} > 0 \quad \text{or} \quad \Im\gamma_p^{(a)} = 0,\ k\gamma_p^{(a)} \ge 0, \tag{4.18}$$

where $\{\lambda_p^{(a)}, \tilde{\Pi}_p\}$ is the complete system of eigenvalues and orthogonal and normalized eigenfunctions (in L_2) of the two-dimensional Laplace operator $-\Delta$ in a rectangle, with the boundary conditions

$$\left.\tilde{\Pi}_p\right|_{x_1=\pm\frac{a}{2}} = \left.\frac{\partial\tilde{\Pi}_p}{\partial x_2}\right|_{x_2=\pm\frac{b}{2}} = 0,$$

where the coefficients $T_p = T_p(y)$ in (4.17) do not depend on x and are uniformly bounded with respect to variables p and y in every bounded subdomain $\overline{T}^+$.

The Green's function G_b^T satisfies ($y \in T^+$ is fixed) the Helmholtz equation

$$\Delta G_b^T + k^2 G_b^T = -\delta(\mathrm{x} - y), \quad \mathrm{x} \in T^+, \tag{4.19}$$

the boundary conditions

$$\left.\frac{\partial G_b^T}{\partial \mathrm{x}_3}\right|_{\mathrm{x}_3=0} = \left.\frac{\partial G_b^T}{\partial \mathrm{x}_1}\right|_{\mathrm{x}_1=\pm\frac{a}{2}} = \left. G_b^T\right|_{\mathrm{x}_2=\pm\frac{b}{2}} = 0, \tag{4.20}$$

and the conditions at infinity: the Green's function can be represented, for $|\mathrm{x}_3| \geq C_0$ and a certain $C_0 > 0$, in the form of the series

$$G_b^T = \sum_p R_p e^{i\gamma_p^{(b)}|\mathrm{x}_3|}\tilde{\Psi}_p(\mathrm{x}_1, \mathrm{x}_2), \tag{4.21}$$

$$\gamma_p^{(b)} = \sqrt{k^2 - \lambda_p^{(b)}}, \ \Im\gamma_p^{(b)} > 0 \quad \text{or} \quad \Im\gamma_p^{(b)} = 0, \ k\gamma_p^{(b)} \geq 0, \tag{4.22}$$

where $\{\lambda_p^{(b)}, \ \tilde{\Psi}_p\}$ is the complete system of eigenvalues and orthogonal and normalized eigenfunctions (in L_2) of the two-dimensional Laplace operator $-\Delta$ in a rectangle, with the boundary conditions

$$\left.\frac{\partial \tilde{\Psi}_p}{\partial \mathrm{x}_1}\right|_{\mathrm{x}_1=\pm\frac{a}{2}} = \left.\tilde{\Psi}_p\right|_{\mathrm{x}_2=\pm\frac{b}{2}} = 0,$$

where the coefficients $R_p = R_p(y)$ in (4.21) do not depend on x and are uniformly bounded with respect to p and y in every bounded subdomain $\overline{T}^+$.

The Green's functions have the form

$$G_a^T = \frac{2i}{ab}\sum_{n=1}^{\infty}\sum_{m=0}^{\infty}\frac{f_{nm}^1(\mathrm{x}')f_{nm}^1(y')}{(1+\delta_{0m})\gamma_{nm}}\left(e^{i\gamma_{nm}|\mathrm{x}_3-y_3|} + e^{i\gamma_{nm}|\mathrm{x}_3+y_3|}\right) \tag{4.23}$$

and

$$G_b^T = \frac{2i}{ab}\sum_{n=0}^{\infty}\sum_{m=1}^{\infty}\frac{f_{nm}^2(\mathrm{x}')f_{nm}^2(y')}{(1+\delta_{0n})\gamma_{nm}}\left(e^{i\gamma_{nm}|\mathrm{x}_3-y_3|} + e^{i\gamma_{nm}|\mathrm{x}_3+y_3|}\right), \tag{4.24}$$

where

$$\gamma_{nm} = \sqrt{k^2 - \lambda_{nm}}, \ \Im\gamma_{nm} > 0 \quad \text{or} \quad \Im\gamma_{nm} = 0, \ k\gamma_{nm} \geq 0,$$

$$\lambda_{nm} = \left(\frac{\pi n}{a}\right)^2 + \left(\frac{\pi m}{b}\right)^2,$$

$$f_{nm}^1(z') = \sin\frac{\pi n}{a}\left(z_1 + \frac{a}{2}\right)\cos\frac{\pi m}{b}\left(z_2 + \frac{b}{2}\right),$$

$$f_{nm}^2(z') = \cos\frac{\pi n}{a}\left(z_1 + \frac{a}{2}\right)\sin\frac{\pi m}{b}\left(z_2 + \frac{b}{2}\right), \quad z' = (z_1, z_2).$$

The method for construction of the Green's function is described, together with their basic properties, in Ref. [43]. Note that series (4.23) and (4.24) converge exponentially when $|\mathrm{x}_3| \neq |y_3|$.

In order to investigate the traces of the Green's functions at $\mathrm{x}_3 = 0$ and $y_3 = 0$, we will separate their singularities for $|\mathrm{x}' - y'| \to 0$.

Consider the integral

$$I_{nm} = \int\limits_{\Delta_{nm}} f(\xi) e^{i\mathrm{x}'\cdot\xi}\, d\xi, \quad f(\xi) = (\xi^2 - k^2)^{-1/2},$$

over the rectangle

$$\Delta_{nm} = \{\xi : |\xi_1 - h_1 n| < h_1, |\xi_2 - h_2 m| < h_2\}$$

(the correct choice of the branch of the square root is described in Section 4.3):

$$I_{nm} = \int\limits_{\Delta_{nm}} L_{nm}(\xi) e^{i\mathrm{x}'\cdot\xi}\, d\xi + \int\limits_{\Delta_{nm}} r_{nm}(\xi) e^{i\mathrm{x}'\cdot\xi}\, d\xi$$

$$= \sum_{l,j=-1}^{1} C_{lj} \frac{\exp\left(i\mathrm{x}_1 h_1(n+l) + i\mathrm{x}_2 h_2(m+j)\right)}{(h_1^2(n+l)^2 + h_2^2(m+j)^2 - k^2)^{1/2}} + R_{nm}(\mathrm{x}'),$$

where $L_{nm}(\xi)$ is the interpolation polynomial of the second degree for the function $f(\xi)$ with the nodes $h_1(n+l)$, $h_2(m+j)$ $(l,j = -1,0,1)$, and $r_{nm} = f - L_{nm}$. Coefficients C_{lj} are expressed by the formulas

$$C_{lj} = \tfrac{1}{\mathrm{x}_1\mathrm{x}_2} C^{(l)}(\mathrm{x}_1 h_1) C^{(j)}(\mathrm{x}_2 h_2),$$
$$C^{(j)} = \tfrac{(-1)^j}{2-\delta_{0j}} \exp(ijt)\Theta_j(t),$$

where

$$\Theta_1(t) = 2\left(1 - \frac{2}{t^2} - \frac{i}{t}\right)\sin t + 2\left(\frac{2}{t} + i\right)\cos t,$$

$$\Theta_2(t) = \frac{4}{t}\cos t - \frac{4}{t^2}\sin t, \quad \Theta_3(t) = \overline{\Theta_1(t)},$$

and C_{lj} do not depend on n and m. In the above formulas, we have assumed that

$$k^2 \neq h_1^2 n^2 + h_2^2 m^2, \quad n, m \in \mathrm{Z}.$$

For the remainder $R_{nm}(\mathrm{x}')$, we have the estimates

$$|R_{nm}(\mathrm{x}')| \le C(n^2+m^2)^{-2},$$

$$\left|\frac{\partial}{\partial \mathrm{x}_i} R_{nm}(\mathrm{x}')\right| \le C(n^2+m^2)^{-3/2} \quad (i = 1,2), \tag{4.25}$$

which hold for large n and m. Indeed, $r_{nm}(\xi)$ can be represented through the third derivatives of $f(\xi)$, which decay asymptotically as $O(|\xi|^{-4})$ for $|\xi| \to \infty$. Estimates (4.25) are uniform with respect to x'.

Calculating integrals I_{nm}, we obtain

$$\sum_{n,m=-\infty}^{\infty} = 4\int\limits_{\mathrm{R}^2} e^{i\mathrm{x}'\cdot\xi} f(\xi)\, d\xi = 8\pi \frac{e^{ik|\mathrm{x}'|}}{|\mathrm{x}'|}.$$

Here, the sum of the series is understood as a limit of partial spherical sums,

$$\sum_{n,m=-\infty}^{\infty} := \lim_{R\to\infty} \sum_{n,m:\Delta_{nm}\subset B_R},$$

where B_R is the circle of the radius R and the center at the origin. Thus, we obtain

$$\alpha(\mathrm{x}')H(\mathrm{x}') + R(\mathrm{x}') = 8\pi \frac{e^{ik|\mathrm{x}'|}}{|\mathrm{x}'|}, \tag{4.26}$$

where

$$\alpha(\mathrm{x}') = \sum_{l,j=-1}^{1} C_{lj} = 64 \prod_{i=1}^{2} \frac{1}{h_i \mathrm{x}_i^2} \sin^2 \frac{\mathrm{x}_i h_i}{2} \left(\frac{\sin \mathrm{x}_i h_i}{\mathrm{x}_i h_i} + \sin^2 \frac{\mathrm{x}_i h_i}{2} \right),$$

$$H(\mathrm{x}') = \sum_{n,m=-\infty}^{\infty} \exp(i\mathrm{x}_1 h_1 n + i\mathrm{x}_2 h_2 m)(h_1^2 n^2 + h_2^2 m^2 - k^2)^{-1/2},$$

$$R(\mathrm{x}') = \sum_{n,m=-\infty}^{\infty} R_{nm}(\mathrm{x}'). \tag{4.27}$$

From (4.26), we find

$$H(\mathrm{x}') = \alpha^{-1}(\mathrm{x}') \left(8\pi |\mathrm{x}'|^{-1} \exp(ik|\mathrm{x}'|) - R(\mathrm{x}') \right), \tag{4.28}$$

and, consequently,

$$H(\mathrm{x}') = \frac{2\pi}{h_1 h_2} \frac{e^{ik|\mathrm{x}'|}}{|\mathrm{x}'|} + \tilde{R}(\mathrm{x}'), \quad \mathrm{x}' \in \Pi_h \tag{4.29}$$

for $\mathrm{x}' \in \Pi_h = \{\mathrm{x}' : |\mathrm{x}_i| < 2\pi/h_i,\ i = 1, 2\}$. From (4.25)–(4.29), it follows that $\tilde{R}(\mathrm{x}') \in C^1(K)$ for every compact set

$$K \subset \{\mathrm{x}' : |\mathrm{x}_1| < 2\pi/h_1,\ |\mathrm{x}_2| < 2\pi/h_2\}.$$

Then,

$$\sum_{n,m \geq 0} \frac{\cos(\mathrm{x}_1 h_1 n)}{1 + \delta_{0n}} \frac{\cos(\mathrm{x}_2 h_2 m)}{1 + \delta_{0m}} \frac{1}{(h_1^2 n^2 + h_2^2 m^2 - k^2)^{1/2}} = \frac{H(\mathrm{x}')}{4}. \tag{4.30}$$

$H(\mathrm{x}')$ is periodic function of x_i with the period $2\pi/h_i$; therefore, extracting or adding the period, we can reduce the procedure of separating singularities as $\mathrm{x}_i \to \pm 2\pi/h_i$ to the case specified by formula (4.29). Then, from (4.30), we obtain the expression for the singularities the traces of the Green's functions $G_{a,b}^T(\mathrm{x}', y')$ in the rectangle $\Pi = \{\mathrm{x}' : |\mathrm{x}_1| \leq a/2,\ |\mathrm{x}_2| < b/2\}$:

$$G_a^T \Big|_{\mathrm{x}_3 = y_3 = 0} = \frac{1}{2\pi} \sum_{n,m=-1}^{1} (-1)^n \frac{e^{ik\rho_{nm}}}{\rho_{nm}} + g_a(\mathrm{x}', y'), \tag{4.31}$$

$$G_b^T \Big|_{\mathrm{x}_3 = y_3 = 0} = \frac{1}{2\pi} \sum_{n,m=-1}^{1} (-1)^m \frac{e^{ik\rho_{nm}}}{\rho_{nm}} + g_b(\mathrm{x}', y'), \tag{4.32}$$

where $\mathrm{x}', y' \in \overline{\Pi}$,

$$\rho_{nm}^2 = (\mathrm{x}_1 - (-1)^n y_1 - na)^2 + (\mathrm{x}_2 - (-1)^m y_2 - mb)^2$$

and functions $g_{a,b} \in C^1(\overline{\Pi})$. It should be noted that expressions (4.31) and (4.32) enable one to separate singularities completely in the closed rectangle $\overline{\Pi}$. However, the remainder

$g_{a,b}$ belongs to $C^1(\overline{\Pi})$, but below, in Lemma 8, we will prove that $g_{a,b} \in C^\infty(\overline{\Pi})$. If $\overline{\Omega} \subset \Pi$, then, we have $\partial\Omega \bigcap \partial\Pi = \emptyset$ and (4.31) yields

$$G_a^T(\mathrm{x}', y') = \frac{1}{2\pi} \frac{e^{ik|\mathrm{x}'-y'|}}{|\mathrm{x}'-y'|} + \tilde{g}_a(\mathrm{x}', y'), \tag{4.33}$$

$$G_b^T(\mathrm{x}', y') = \frac{1}{2\pi} \frac{e^{ik|\mathrm{x}'-y'|}}{|\mathrm{x}'-y'|} + \tilde{g}_b(\mathrm{x}', y'), \quad \tilde{g}_{a,b} \in C^1(\overline{\Omega}). \tag{4.34}$$

Convergence of the series for the Green's function and for the function $H(\mathrm{x}')$ is conventional and is studied in Refs. [20, 21, 22]. Note that by choosing the interpolation polynomial $L_{nm}(\xi)$ of a higher degree, one can improve the convergence of the series for the remainder $R(\mathrm{x}')$ and to attain, for arbitrary N, the estimate $R_{nm}(\mathrm{x}') = O((n^2+m^2)^{-N})$.

Consider the dependence of the Green's functions on parameter k. Denote by $\Lambda_{a,b}(T)$ the set of the points, at which the Green's functions $G_{a,b}^T$ are not defined:

$$\Lambda_a(T) = \left\{ k : k^2 = \left(\frac{\pi n}{a}\right)^2 + \left(\frac{\pi m}{b}\right)^2, n \geq 1, m \geq 0 \right\},$$

$$\Lambda_b(T) = \left\{ k : k^2 = \left(\frac{\pi n}{a}\right)^2 + \left(\frac{\pi m}{b}\right)^2, n \geq 0, m \geq 1 \right\}.$$

The branch of the square root for $\gamma_p^{(a,b)}$ is chosen according to (4.18) and (4.22),

$$\gamma = \sqrt{k^2 - \lambda} = \frac{1}{\sqrt{2}} \left(sign\Re k \sqrt{|k^2 - \lambda| + \Re\left(k^2 - \lambda\right)} + i\sqrt{|k^2 - \lambda| - \Re\left(k^2 - \lambda\right)} \right), \tag{4.35}$$

where $\gamma = \gamma_p^{(a,b)}$ and $\lambda = \lambda_p^{(a,b)}$.

Remark 6 *The function $\gamma = \sqrt{k^2 - \lambda}$ may be continued analytically to the complex plane k cut along the lower semi-circumference that joins the points $\sqrt{\lambda}$ and $-\sqrt{\lambda}$.*

According to the conventional terminology, conditions (4.17) and (4.18) and (4.21) and (4.22) mean that, for $\Im k = 0$ and $k > 0$, the Green's functions can be represented as superpositions of radiating waves, so that a finite number of them (for $\Im \gamma_p = 0$ and $\Re \gamma_p > 0$) are propagating waves, and the rest (for $\Im \gamma_p > 0$) are decaying waves.

Assume that $|\mathrm{x}_3| \neq |y_3|$, $\mathrm{x}', y' \in \overline{\Pi}$. Then, obviously, the Green's functions $G_{a,b}^T = G_{a,b}^T(\mathrm{x}, y; k)$ are analytical in C_+ with respect to k and continuous at the points $\overline{\mathrm{C}}_+ \setminus \Lambda_{a,b}(T)$ because the series converge exponentially. The same result is valid for the traces of the Green's functions $G_{a,b}^T(\mathrm{x}', y'; k)$ at $\mathrm{x}' \neq y'$, $\mathrm{x}', y' \in \overline{\Omega} \subset \Pi$: they are analytical in C_+ and continuous in $\overline{\mathrm{C}}_+ \setminus \Lambda_{a,b}(T)$. In order to verify the latter property, it is sufficient to separate singularities using formulas (4.33) and (4.34) and to differentiate the functions with respect to k. Since $\alpha(\mathrm{x}')$ does not depend on k and the series for $R(\mathrm{x}'; k)$ and $\partial R(\mathrm{x}'; k)/\partial k$ converge uniformly in the rectangle $\overline{\Pi}$ because their terms behave as $O((n^2 + m^2)^{-2})$ and $O((n^2 + m^2)^{-3/2})$, respectively, we may complete the proof.

For the Green's functions of the domain

$$T^- = \{\mathrm{x} : |\mathrm{x}_1| < a/2,\ |\mathrm{x}_2| < b/2,\ \mathrm{x}_3 < 0\},$$

which admit representations similar to (4.23) and (4.24), but for $x, y \in T^-$, all the above formulas and properties remain valid if to replace T^+ by T^-. Note that this similarity holds because all expressions in formulas (4.17) and (4.21) contain the absolute value of x_3.

Let us prove the general statement concerning singularities of the Green's functions G^U and $G^T_{a,b}$, which differs from traditional theorems. In fact, we separate singularities not only at the internal points of a domain, but also at the boundary points.

Lemma 8 *The Green's functions G^U and $G^T_{a,b}$ admit the following representations:*

$$G^U(x,y) = \frac{e^{ik|x-y|}}{|x-y|} + \frac{e^{ik|x-y^*|}}{|x-y^*|} + \frac{e^{ik|x-y^*+2e_3|}}{|x-y^*+2e_3|} + V^U(x,y), \tag{4.36}$$

where $y^* = (y_1, y_2, -y_3)$ and $V^U(x,y) \in C^\infty(\overline{U} \times \overline{U})$,

$$G^T_a(x,y) = \frac{1}{4\pi} \sum_{n,m=-1}^{1} (-1)^n \left(\frac{e^{ikr_{nm}}}{r_{nm}} + \frac{e^{ikr^*_{nm}}}{r^*_{nm}} \right) + V^T_a(x,y), \tag{4.37}$$

and

$$G^T_b(x,y) = \frac{1}{4\pi} \sum_{n,m=-1}^{1} (-1)^m \left(\frac{e^{ikr_{nm}}}{r_{nm}} + \frac{e^{ikr^*_{nm}}}{r^*_{nm}} \right) + V^T_b(x,y), \tag{4.38}$$

where

$$r_{nm}(x,y) = \Big((x_1 - (-1)^n y_1 - na)^2$$
$$+ (x_2 - (-1)^m y_2 - mb)^2 + (x_3 - y_3)^2 \Big)^{1/2},$$
$$r^*_{nm}(x,y) = r_{nm}(x,y^*), \quad V^T_{a,b} \in C^\infty(\overline{T^+} \times \overline{T^+}).$$

□ We prove this lemma applying the principle of the source reflection [4, p. 426]. Let us begin with representation (4.36). Denote by $G^U_0(x,y)$ the Green's function of the Helmholtz equation (4.6) of the "doubled" layer $U_0 = \{x : |x_3| < 1\}$, which is obtained as a result of symmetric reflection with respect to the plane $x_3 = 0$ with the boundary conditions of the second kind at $x_3 = \pm 1$ and the Sveshnikov–Werner conditions at infinity. Here, we will not use the explicit expression, which can be easily obtained from (4.11) or (4.12)]. Using the source reflection, we obtain

$$G^U(x,y) = G^U_0(x,y) + G^U_0(x,y^*), \quad x,y \in U.$$

It is known [4] that

$$G^U_0(x,y) = \frac{1}{4\pi} \frac{e^{ik|x-y|}}{|x-y|} + V^U_0(x,y), \quad x,y \in U_0,$$

where $V^U_0(x,y) \in C^\infty(U_0)$ as a solution to the homogeneous Helmholtz equation in the domain U_0. Therefore,

$$G^U(x,y) = \frac{1}{4\pi} \frac{e^{ik|x-y|}}{|x-y|} + \frac{1}{4\pi} \frac{e^{ik|x-y^*|}}{|x-y^*|} + V^U_{(1)}(x,y),$$

where the function

$$V^U_{(1)} \in C^\infty \left((U \bigcup \{x_3 = 0\}) \times (U \bigcup \{x_3 = 0\}) \right)$$

is infinitely differentiable not only in the open domain U but also in the plane $x_3 = 0$. Performing a similar reflection of U with respect to the plane $x_3 = -1$, we find

$$G^U(x, y) = \frac{1}{4\pi}\frac{e^{ik|x-y|}}{|x-y|} + \frac{1}{4\pi}\frac{e^{ik|x-y^*+2e_3|}}{|x-y^*+2e_3|} + V^U_{(2)}(x, y),$$

where

$$V^U_{(2)} \in C^\infty\left((U\bigcup\{x_3 = -1\}) \times (U\bigcup\{x_3 = -1\})\right).$$

Combining these representations, we obtain expression (4.36).

Let us prove representation (4.37). Note that formula (4.38) can be verified in a similar manner. Denote by $G^T_0(x, y)$ the Green's function of the Helmholtz equation (4.15) of the infinite rectangular cylinder

$$T_0 = \{x : |x_1 - a/2| < a, |x_2 - b/2| < b\},$$

which is obtained by adding the symmetric reflection of T^+ in the planes $x_3 = 0$, $x_1 = a/2$, and $x_2 = b/2$ with the boundary conditions of the first kind at $|x_1 - a/2| = a$, the boundary conditions of the second kind at $|x_2 - b/2| = b$, and the Sveshnikov conditions at infinity. Again, we will not use the corresponding explicit expression, which can be derived from (4.23). Using the source reflection and taking into account the boundary conditions, we obtain

$$\begin{aligned} G^T(x, y) &= G^T_0(x, y) - G^T_0(x, y') + G^T_0(x, y'') - G^T_0(x, y''') \\ &\quad + G^T_0(x, y^*) - G^T_0(x, y'^*) + G^T_0(x, y''^*) - G^T_0(x, y'''^*), \end{aligned}$$

where $x, y \in T^+$,

$$y' = (a - y_1, y_2, y_3),\ y'' = (y_1, b - y_2, y_3),\ y''' = (a - y_1, b - y_2, y_3).$$

On the other hand,

$$G^T_0(x, y) = \frac{1}{4\pi}\frac{e^{ik|x-y|}}{|x-y|} + V^T_0(x, y), \quad x, y \in T_0,$$

where $V^T_0(x, y) \in C^\infty(T_0)$. From the latter formula, we obtain

$$G^T_a = \frac{1}{4\pi}\sum_{n,m=0}^{1}(-1)^n\left(\frac{e^{ikr_{nm}}}{r_{nm}} + \frac{e^{ikr^*_{nm}}}{r^*_{nm}}\right) + V^T_{(1)}(x, y),$$

where the function

$$V^T_{(1)} \in C^\infty\left((\overline{T^+}\bigcap T_0) \times (\overline{T^+}\bigcap T_0)\right)$$

is infinitely differentiable not only in the open domain T^+, but at the points that belong to the part of the boundary

$$\partial T^+ \setminus \left(\{x_1 = -a/2\}\bigcup\{x_2 = -b/2\}\right).$$

Three more versions of the reflections of T^+ in the planes $x_3 = 0$, $x_1 = -a/2$, $x_2 = b/2$, $x_3 = 0$, $x_1 = a/2$, $x_2 = -b/2$, and $x_3 = 0$, $x_1 = -a/2$, $x_2 = -b/2$ are considered in the similar manner, which enables us to separate all singularities of the Green's function G^T_a on other parts of the boundary ∂T^+. Combining all representations, we obtain expression (4.37). □

The next statement directly follows from Lemma 8.

Corollary 5 *The traces of the Green's functions at $x_3 = 0$ and $y_3 = 0$ can be represented as*

$$G^U(x', y') = \frac{1}{2\pi} \frac{e^{ik|x'-y'|}}{|x'-y'|} + v^U(x', y'), \quad x', y' \in R^2 \tag{4.39}$$

and

$$G^T_{a,b}(x', y') = \frac{1}{2\pi} \frac{e^{ik|x'-y'|}}{|x'-y'|} + v^T_{a,b}(x', y'), \quad x', y' \in \Pi, \tag{4.40}$$

where $v^U \in C^\infty(R^2 \times R^2)$ and $v^T_{a,b} \in C^\infty(\Pi \times \Pi)$.

Formulas (4.39) and (4.40) enable one to perform a complete separation of singularities on every compact set $\overline{\Omega}$ in R^2 or Π. Thus, the results of Lemma 8 and Corollary 5 turn out to be stronger, for functions $G^T_{a,b}(x', y')$, than the statements given by formulas (4.33) and (4.34). However, unlike the first method, one cannot use Lemma 8 as a practical algorithm for separating singularities.

4.2 Vector integrodifferential equation on Ω

In this section, we consider the Fredholm property and solvability of the basic vector integrodifferential equation on Ω, to which all the diffraction problems are reduced.

Let k_1, k_2, μ_1, and μ_2 be the complex numbers such that $\Im k_j \geq 0$, $k_j \neq 0$, $\Im \mu_j \geq 0$, and $\Re \mu_j > 0$, $j = 1, 2$.

Let $\Omega \subset R^2$ be the planar domain with the piece-wise smooth boundary $\Gamma = \partial\Omega$ formed by a finite number of simple C^∞-arcs that join at nonzero angles. Consider the integrodifferential equation on Ω

$$\sum_{j=1}^{2} \left(\mu_j^{-1} \left(grad\, A_j\, (div\, u) + k_j^2\, A_j u \right) + B_j u \right) = f, \quad x \in \Omega, \tag{4.41}$$

where $x = (x_1, x_2)$, $u = (u^1, u^2)^T \in W(\overline{\Omega})$,

$$(A_j u)(x) = \int_\Omega \frac{e^{ik_j|x-y|}}{|x-y|} u(y)\, dy, \tag{4.42}$$

$$(B_j u)(x) = \int_\Omega b_j(x, y) u(y)\, dy, \tag{4.43}$$

and vector-function f and matrix-functions b_j satisfy the condition

$$b_j(x, y) \in C^\infty(\overline{\Omega} \times \overline{\Omega}), \quad f(x) \in C^\infty(\overline{\Omega}). \tag{4.44}$$

Equation (4.41) is similar to (2.19), which has been considered in Sections 2.4–2.6. According to (2.32), we rewrite equation (4.41) as a PDO

$$\sum_{j=1}^{2} \int \mu_j^{-1} a_j(\xi) \left(-\xi(\xi \cdot \hat{u}(\xi)) + k_j^2 \hat{u}(\xi) \right) e^{ix\cdot\xi}\, d\xi$$

$$+ \int_\Omega b(x, y)\, u(y)\, dy = f(x), \quad x \in \Omega, \tag{4.45}$$

or, in the operator form, as

$$Su \equiv Lu + Bu = f, \quad x \in \Omega. \tag{4.46}$$

Here, $\hat{u}$ is the Fourier transform of the vector-function u and $b(x,y) = b_1(x,y) + b_2(x,y)$. Operator S is considered in the spaces $S : W \to W'$, where $W = W(\overline{\Omega})$ and $W' = W'(\Omega)$. For the symbol $a_j(\xi)$, according to (2.25) and (2.28), we have

$$a_j(\xi) = (\xi^2 - k_j^2)^{-1/2}, \quad \Im k_j \geq 0, \tag{4.47}$$

or

$$a_j(\xi) = \frac{1}{\langle\xi\rangle} + \frac{k_j^2+1}{2\langle\xi\rangle^3} + \hat{g}_j(\xi), \tag{4.48}$$
$$\hat{g}_j \in C^\infty(\mathrm{R}^2), \quad \hat{g}_j = O(|\xi|^{-7/2}),$$

where $\hat{g}_j(\xi)$ is defined by formula (2.27) at $k = k_j$. Just as in Chapter 1, formulas (4.47) and (4.48) define one and the same bounded operator $L : W \to W'$. $B : W \to W'$, which is an operator with an infinitely smooth kernel, is bounded and compact on this pair of spaces. Thus, $S : W \to W'$ is a bounded operator.

Equation (4.46) was considered in Chapter 2 at $k_1 = k_2$, $\mu_1 = \mu_2$, and $B = 0$. Since operator B is compact, S will be a Fredholm operator with the zero index because this property holds for operator L. For this operator, we obtain, by virtue of (2.38), the space decomposition

$$\begin{pmatrix} \left(\frac{k_1^2}{\mu_1} + \frac{k_2^2}{\mu_2}\right)(1-\Delta)^{-1/2} + K_1 & 0 \\ 0 & -\left(\frac{1}{\mu_1} + \frac{1}{\mu_2}\right)(1-\Delta)^{1/2} + K_2 \end{pmatrix} : \begin{pmatrix} W_1 \\ W_2 \end{pmatrix} \to \begin{pmatrix} W^1 \\ W^2 \end{pmatrix}, \tag{4.49}$$

with compact operators $K_j : W_j \to W^j$ $(j = 1,2)$. Note that a similar decomposition of the complete operator S is not generally diagonal. Therefore, unlike the case corresponding to equations (2.43) and (2.44), (4.46) does not split into two independent equations.

Theorem 11 *If $\mu_1^{-1}k_1^2 + \mu_2^{-1}k_2^2 \neq 0$ and $\mu_1^{-1} + \mu_2^{-1} \neq 0$, then operator $S : W \to W'$ is Fredholm and* $\operatorname{ind} S = 0$.

□ The proof follows from decomposition (4.49) if to take into account that $(1-\Delta)^{-1/2} : W_1 \to W^1$ and $(1-\Delta)^{1/2} : W_2 \to W^2$ are continuously invertible operators. □

Corollary 6 *If the homogeneous equation $Su = 0$, $u \in W$, has only a trivial solution, then equation (4.41) is uniquely solvable for arbitrary right-hand side $f \in W'$, in particular, for $f \in C^\infty(\overline{\Omega})$.*

Since $Bu \in C^\infty(\overline{\Omega})$ for $u \in W$, Statement 22 implies a similar corollary.

Statement 26 *If $u \in W$ is a solution to equation (4.41) with a smooth right-hand side $f \in C^\infty(\overline{\Omega})$, then $u \in C^\infty(\Omega)$.*

We note that here, all the statements of Section 2.6 are valid that describe smoothness of solution u in the vicinity of the boundary Γ and its corner points.

We will prove one simple statement concerning the dependence of solutions to equation (4.41) on parameters k_1 and k_2. We fix μ_1 and μ_2 and set $S = S(k_1, k_2)$ and $f = f(k_1, k_2)$. Let k_j^0 with $\Im k_j^0 = 0$ be a point on the real axis and $Q_j = \{k_j : |k_j - k_j^0| < \delta,\ \Im k_j \geq 0\}$, $j = 1, 2$. Then the following statement is valid.

Statement 27 *If $f(k_1, k_2) \xrightarrow{W'} f(k_1^0, k_2^0)$, operator-valued function $S(k_1, k_2) : W \to W'$ is continuous in $Q_1 \times Q_2$ with respect to the norm, and operator $S(k_1^0, k_2^0)$ is continuously invertible, then*

$$u(k_1, k_2) \xrightarrow{W} u(k_1^0, k_2^0)$$

for $k_1 \to k_1^0$ and $k_2 \to k_2^0$, where $u(k_1, k_2)$ and $u(k_1^0, k_2^0)$ are solutions to equation (4.41) at k_1 and k_2 and k_1^0 and k_2^0, respectively, where $\Im k_j^0 = 0$, $\Im k_j > 0$, and $k_j \in Q_j$ $(j = 1, 2)$.

□ In order to prove this statement, we apply the identity

$$S^{-1}(k_1, k_2) = S^{-1}(k_1^0, k_2^0)\left(I + \left(S(k_1, k_2) - S(k_1^0, k_2^0)\right) S^{-1}(k_1^0, k_2^0)\right)^{-1},$$

which holds for sufficiently small δ and yields a continuous dependence of $S^{-1}(k_1, k_2)$ on parameters in $Q_1 \times Q_2$, and, consequently, a continuous dependence of $u(k_1, k_2) = S^{-1}(k_1, k_2) f(k_1, k_2)$ for $k_1 \to k_1^0$ and $k_2 \to k_2^0$. □

4.3 Diffraction by a hole in a planar screen

Consider the problem of diffraction of a monochromatic field E^0, H^0 by a hole in a perfectly conducting plane. We assume that the plane is infinitely thin and the hole $\Omega \subset \mathrm{R}^2 = \{x_3 = 0\} \subset \mathrm{R}^3$ is a bounded domain with the piece-wise smooth boundary $\Gamma = \partial\Omega$ formed by a finite number of simple C^∞-arcs that join at nonzero angles. The incident field E^0, H^0 is a solution to the Maxwell equations with the boundary condition

$$E_\tau^0|_{x_3=0} = 0 \tag{4.50}$$

and is generated by sources situated outside $\overline{\Omega}$. Therefore,

$$H_\tau^0|_\Omega \in C^\infty(\overline{\Omega}). \tag{4.51}$$

In R^3_-, the field E^0, H^0 vanishes identically.

We will assume that the media in R^3_+ and R^3_- have the constant permittivity, permeability, and conductivity $\varepsilon_1, \mu_1, \sigma_1$ and $\varepsilon_2, \mu_2, \sigma_2$, respectively, such that

$$\Im \varepsilon_j \geq 0,\ \Re \varepsilon_j > 0,\ \Im \mu_j \geq 0,\ \Re \mu_j > 0,\ k_j^2 = \varepsilon_j \mu_j \omega^2,\ \Im k_j \geq 0,$$

where $k_j \neq 0$ and $\omega > 0$ is the circular frequency. Dielectrics with and without absorption, when

$$\Im \mu_j = 0,\ \mu_j > 0,\ \varepsilon_j = \varepsilon_j^0 + i\sigma_j\omega^{-1},\ \sigma_j \geq 0,\ \varepsilon_j^0 > 0,$$

constitute a particular case of this model.

The problem of diffraction by a hole in a planar screen is reduced to determination of the scattered field

$$E, H \in C^2(\mathrm{R}^3_+ \bigcup \mathrm{R}^3_-) \bigcap_{\delta>0} C(\overline{\mathrm{R}^3_+} \setminus \Gamma_\delta) \bigcap_{\delta>0} C(\overline{\mathrm{R}^3_-} \setminus \Gamma_\delta), \tag{4.52}$$

that satisfies the Maxwell equations

$$\begin{aligned} Rot\, H &= -i\omega\varepsilon\, E, \\ Rot\, E &= i\omega\mu\, H, \quad \mathrm{x} \in \mathrm{R}^3_+ \bigcup \mathrm{R}^3_-, \end{aligned} \tag{4.53}$$

where $\varepsilon = \varepsilon_1$ and $\mu = \mu_1$ in R^3_+ and $\varepsilon = \varepsilon_2$ and $\mu = \mu_2$ in R^3_-, the boundary condition

$$E_\tau|_\Sigma = 0 \tag{4.54}$$

for the tangential field components on the surface $\Sigma = \mathrm{R}^2 \setminus \overline{\Omega} \subset \{\mathrm{x} : \mathrm{x}_3 = 0\}$ of the perfectly conducting screen, the conjugation conditions on the interface

$$[E_\tau]_\Omega = 0, \tag{4.55}$$

$$[H_\tau]_\Omega = -[H^0_\tau]_\Omega, \tag{4.56}$$

the condition of the finiteness of energy in every bounded volume

$$E, H \in L^2_{loc}(\mathrm{R}^3), \tag{4.57}$$

and the conditions at infinity

$$E, H = o(r^{-1}), \quad r = |\mathrm{x}| \to \infty, \tag{4.58}$$

for $\Im\varepsilon > 0$ or $\Im\mu > 0$, and

$$E \times e_r + (\mu/\varepsilon)^{1/2} H = o(r^{-1}), \quad H = O(r^{-1}), \quad r \to \infty, \tag{4.59}$$

for $\Im\varepsilon = 0$, $\Im\mu = 0$, $\varepsilon > 0$, and $\mu > 0$, uniformly along all directions $e_r = \mathrm{x}/|\mathrm{x}|$, where $\mathrm{x} \in \mathrm{R}^3_+$ or $\mathrm{x} \in \mathrm{R}^3_-$. The solution to problem (4.52)–(4.59) will be called quasiclassical.

Before to prove uniqueness of the solution to diffraction problem (4.52)–(4.59), we establish the validity of the following statement.

Lemma 9 *Let the function* $u \in C^2(\mathrm{R}^3_+ \setminus \overline{B_R}) \bigcap C^1(\overline{\mathrm{R}^3_+} \setminus B_R)$, *where* $B_R = \{\mathrm{x} : |\mathrm{x}| < R\}$, *be a solution to the homogeneous Helmholtz equation*

$$\Delta u + k^2 u = 0 \quad in \quad \mathrm{R}^3_+ \setminus \overline{B_R}$$

for $\Im k = 0$ *and* $k \neq 0$ *with the Dirichlet or the Neumann boundary conditions at* $\mathrm{x}_3 = 0$, $|\mathrm{x}| > R$. *If*

$$\lim_{r\to\infty} \int\limits_{S^+_r} |u(\mathrm{x})|^2\, ds = 0, \quad S^+_r := \{\mathrm{x} : |\mathrm{x}| = r, \mathrm{x}_3 > 0\},$$

then $u \equiv 0$ *in* $\overline{\mathrm{R}^3_+} \setminus B_R$.

□ We set $v(\mathrm{x}) = (sign\, \mathrm{x}_3)\, u(\mathrm{x}_1, \mathrm{x}_2, |\mathrm{x}_3|)$ for the Dirichlet condition and $v(\mathrm{x}) = u(\mathrm{x}_1, \mathrm{x}_2, |\mathrm{x}_3|)$ for the Neumann condition, where $\mathrm{x} \in \mathrm{R}^3$. Then, $[v] = [\partial v/\partial \mathrm{x}_3] = 0$ at $\mathrm{x}_3 = 0$, $|\mathrm{x}| > R$. Therefore, according to Ref. [6], we have

$$\Delta v + k^2 v = 0 \quad \text{in} \quad \mathrm{R}^3 \setminus \overline{B_R}, \quad v \in C^2(\mathrm{R}^3 \setminus B_R),$$

and $\lim\limits_{r\to\infty} \int\limits_{|\mathrm{x}|=r} |v(\mathrm{x})|^2\, ds = 0$, if to take into account the condition of the lemma. Finally, we use the Röllich lemma [23, p. 50] to show that $v \equiv 0$ in $\mathrm{R}^3 \setminus \overline{B_R}$ and $u \equiv 0$ in $\overline{\mathrm{R}^3_+} \setminus B_R$.
□

Lemma 10 *Assume that $E, H \in C^2(D) \bigcap C(\overline{D})$, where*

$$D = \mathrm{R}^3_+ \bigcap B_R(x_0), \quad B_R(x_0) = \{\mathrm{x} : |\mathrm{x} - x_0| < R\},$$

and $x_0 = (\mathrm{x}_1, \mathrm{x}_2, 0)$. If E, H satisfy in D the homogeneous Maxwell equations (4.53) and the homogeneous boundary condition $E_\tau|_\Sigma = 0$ at $\mathrm{x}_3 = 0$ and $\mathrm{x} \in B_R(x_0)$, then $E, H \in C^\infty(\overline{D}_1)$, where $D_1 = \mathrm{R}^3_+ \bigcap B_{R_1}(x_0)$ and $B_{R_1}(x_0) = \{\mathrm{x} : |\mathrm{x} - x_0| < R_1\}$, $R_1 < R$.

□ First of all, it should be noted that all the field components satisfy in D the homogeneous Helmholtz equation with parameter k^2 [27]. Using the boundary condition and the regularity of solutions to elliptic equations [31] (for example, the Schauder estimates), we prove that $E_\tau \in C^m(\overline{D}_1)$ for every $m \geq 1$. The Maxwell equations yield $Div\, E = 0$ in D. Therefore, $\partial E_\nu / \partial \mathrm{x}_3 = 0$ at $\mathrm{x}_3 = 0$ for $\mathrm{x} \in B_R(x_0)$, and again, we can apply the regularity theorem to conclude that $E_\nu \in C^m(\overline{D}_1)$ for every m. Finally, the second equation in (4.53) yields $H \in C^\infty(\overline{D}_1)$. □

Note that Lemmas 9 and 10 remain valid if to replace R^3_+ by R^3_- and S_r^+ by $S_r^- := \{\mathrm{x} : |\mathrm{x}| = r,\ \mathrm{x}_3 < 0\}$.

Theorem 12 *Problem (4.52)–(4.59) has not more than one solution for $\Im \varepsilon \geq 0$, $\Re \varepsilon > 0$, $\Im \mu \geq 0$, $\Re \mu > 0$, and $\omega > 0$*

□ It is sufficient to show that homogeneous problem (4.52)–(4.59) has only a trivial solution. Let $[H_\tau^0]_\Omega \equiv 0$. We choose the value of R so that $\overline{\Omega} \subset B_R$, $B_R = \{\mathrm{x} : |\mathrm{x}| < R\}$, and apply the Lorentz lemma in $B_R \bigcap \mathrm{R}^3_+$ and $B_R \bigcap \mathrm{R}^3_-$. Calculating the sum of the resulting expressions and applying the boundary conditions and the conjugation conditions, we obtain

$$\Re \int\limits_{\partial B_R} (E \times \overline{H}) n \, ds + \omega \int\limits_{B_R} (\Im \varepsilon \, |E|^2 + \Im \mu \, |H|^2) \, d\mathrm{x} = 0, \tag{4.60}$$

where n is the external normal on the sphere ∂B_R. The possibility of applying the Lorentz lemma is justified in Section 2.1.

Let us introduce the notations $D_1 = \mathrm{R}^3_+$, $D_2 = \mathrm{R}^2_-$, $S_1 = S_R^+$, $S_2 = S_R^-$, and

$$J_i = \Re \int\limits_{S_i} (E \times \overline{H}) n \, ds, \quad i = 1, 2.$$

If $\Im(\varepsilon_i + \mu_i) > 0$, then, (4.58) yields $J_i = o(1)$ as $R \to \infty$. For real ε_i and μ_i, we have, according to (4.59),

$$J_i = \left(\frac{\mu_i}{\varepsilon_i}\right)^{1/2} \int\limits_{S_i} |H|^2 \, ds + o(1), \quad R \to \infty.$$

Performing the limiting transition $R \to \infty$ in (4.60), we see that either $E \equiv 0$ and $H \equiv 0$ in D_i or

$$\lim_{R \to \infty} \int\limits_{S_i} |H|^2 \, ds = 0 \quad \text{for} \quad \Im \varepsilon_i = \Im \mu_i = 0.$$

In the second case, one can use Lemma 9 to prove that $H \equiv 0$ (and, consequently, $E \equiv 0$) for $|\mathrm{x}| > R$, $\mathrm{x} \in D_i$, because the components of H_ν and H_τ satisfy, at $\mathrm{x}_3 = 0$, the homogeneous Helmholtz equation with the Dirichlet and the Neumann conditions, respectively. Here, Lemma 9 may be applied because the conditions of Lemma 10 and (4.52) are valid. But the components of E and H are analytical functions in D_i (as solution the homogeneous Helmholtz equation); therefore, $H \equiv 0$ and $E \equiv 0$ in D_i. We may conclude that $E \equiv 0$ and $H \equiv 0$ in R^3. □

We will look for the solution to problem (4.52)–(4.59) in the form

$$E = E_1,\ H = H_1 \quad \text{for} \quad \mathrm{x} \in \mathrm{R}^3_+, \tag{4.61}$$

$$E = E_2,\ H = H_2 \quad \text{for} \quad \mathrm{x} \in \mathrm{R}^3_-, \tag{4.62}$$

$$E_j = \frac{(-1)^{j+1}}{2\pi} \, Rot\, A_j u, \tag{4.63}$$

$$H_j = \frac{(-1)^{j+1}}{2\pi i \omega \mu_j} \left(Grad\, Div\, A_j u + k_j^2 A_j u \right), \tag{4.64}$$

$$A_j u = 2\pi \int\limits_{\Omega} G_j^R(\mathrm{x}, y) u(y)\, dy = \int\limits_{\Omega} \frac{e^{ik_j|\mathrm{x}-y|}}{|\mathrm{x} - y|} u(y)\, dy. \tag{4.65}$$

We will assume that the vector-function $u = (u^1, u^2)^T$ satisfies the condition

$$u,\ div\, u \in C^1(\Omega), \quad u \in W(\overline{\Omega}). \tag{4.66}$$

Representations (4.63)–(4.65) are analyzed, in combination with conditions (4.66), in Section 2.3. In particular, we prove the relationship $Div\, A_j u = A_j(div\, u)$, which yields a modified representation for the field H,

$$H_j = \frac{(-1)^{j+1}}{2\pi i \omega \mu_j} \left(Grad\, A_j(div\, u) + k_j^2 A_j u \right), \tag{4.67}$$

which is equivalent to (4.64). It is clear that fields E and H belong to $C^2(\mathrm{R}^3_+ \cup \mathrm{R}^3_-)$ and that they satisfy the Maxwell equations (4.53), conditions (4.57) that provide the finiteness of energy in every bounded volume, and conditions at infinity (4.58) and (4.59); the constants in equations (4.63) and (4.64) are chosen so that the left-hand sides satisfy equations (4.53) and conditions (4.59). The components H_τ and E_ν defined by formulas (4.63) and (4.67) are continuous up to Ω from each side of the surface (except for the points on Γ). The differentiation and limiting transitions can be performed under the integral sign, and the integrals are understood in the sense of Cauchy. The following limiting relationships are valid:

$$\lim_{\mathrm{x}_3 \to +0} H_\nu = \frac{i}{\omega \mu_1} div\, u, \quad \lim_{\mathrm{x}_3 \to -0} H_\nu = \frac{i}{\omega \mu_2} div\, u,$$

$$\lim_{\mathrm{x}_3 \to \pm 0} E_\tau = u \times e_3, \quad x \in \Omega, \quad \nu = e_3 = (0,0,1). \tag{4.68}$$

Consequently, the conjugation condition (4.55) is fulfilled.

We define H_ν and E_τ on each side of Ω using formulas (4.68). As has been shown in Section 2.3, the limiting transition $\mathrm{x} \to x_0 \in \mathrm{R}^2 \setminus \overline{\Omega}$ and $\mathrm{x}_3 \to \pm 0$ in (4.63) and (4.64) can be performed under the integral sign; therefore, fields E and H are continuous on each side of the plane $\mathrm{x}_3 = 0$ at the points $x_0 \in \mathrm{R}^2 \setminus \overline{\Omega}$, and conditions (4.52) are satisfied completely. In addition to this, calculating E_τ by formula (4.63) at the points $x_0 \in \Sigma$, we prove that condition (4.54) is also valid.

Thus, representation (4.61)–(4.65) guarantees the fulfillment of all conditions of the problem of diffraction by a hole (4.52)–(4.59), except for (4.56), for every u that satisfies (4.66). The conjugation condition (4.56) yields the integrodifferential equation on Ω:

$$\sum_{j=1}^{2} \mu_j^{-1} \left(grad\, A_j\,(div\, u) + k_j^2\, A_j u\right) = f, \quad x \in \Omega, \quad u \in W, \tag{4.69}$$

$$(A_j u)(x) = \int_{\Omega} \frac{e^{ik_j|x-y|}}{|x-y|} u(y)\, dy, \tag{4.70}$$

$$f(x) = -2\pi i\omega [H_\tau^0]_\Omega, \quad f \in C^\infty(\overline{\Omega}), \tag{4.71}$$

that is, equation (4.41) with $b_j = 0$. Applying Theorem 11, it is easy to show that homogeneous equation (4.69) (at $f = 0$) cannot have a nontrivial solution. Otherwise, by virtue of (4.68), formulas (4.61)–(4.65) would give a nontrivial solution to homogeneous diffraction problem (4.52)–(4.59), which is impossible due to Theorem 12. Then, from (4.71) and Statement 26, it follows that every solution $u \in W$ satisfies conditions (4.66). We have proved the following

Theorem 13 *Assume that $\Im\varepsilon \geq 0$, $\Re\varepsilon > 0$, $\Im\mu \geq 0$, $\Re\mu > 0$, and $\omega > 0$. Then, equation (4.69) is uniquely solvable for arbitrary right-hand side $f \in W'$, in particular, for $f \in C^\infty(\overline{\Omega})$. Problem (4.52)–(4.59) has the unique solution, which can be represented in the form (4.61)–(4.65), where u satisfies conditions (4.66) and equation (4.69).*

Now we can prove that the principle of limiting absorption is valid for the problem of diffraction by a hole in a screen.

Statement 28 *Let us fix parameters μ, k^0, and ω such that $\Im\mu_j = 0$, $\mu_j > 0$, $\Im k_j^0 = 0$, $k_j^0 \neq 0$ $(j = 1, 2)$, and $\omega > 0$. Then, if $f(k_1, k_2) \overset{W'}{\to} f(k_1^0, k_2^0)$ for $k_j \to k_j^0$ with $\Im k_j > 0$, we have $u(k_1, k_2) \overset{W}{\to} u(k_1^0, k_2^0)$, $E(k_1, k_2) \to E(k_1^0, k_2^0)$, and $H(k_1, k_2) \to H(k_1^0, k_2^0)$ in $L^2_{loc}(\mathrm{R}^3)$ for $k_j \to k_j^0$ $(j = 1, 2)$, where u is a solution to equation (4.69) and E, H is a solution to problem (4.52)–(4.59), obtained for corresponding values of parameters ε, μ, and k.*

□ The proof follows from the analytical dependence in (4.70) on parameter k_j, Theorem 3, and Statement 27, because the conditions of this statement are valid in this case. The properties of fields E and H can be verified with the help of Theorem 3 and representations (4.61)–(4.65). □

Usually, $\varepsilon_j = \varepsilon_j^0 + i\sigma_j\omega^{-1}$, where $\sigma_j \geq 0$ and $\varepsilon_j^0 > 0$. The principle of limiting absorption means that, for $\sigma_j = 0$, a solution to equation (4.69) and problem (4.52)–(4.59) can be obtained through the limiting transition $\sigma_j \to +0$ $(\sigma_j > 0)$.

4.4 Diffraction by a partially shielded layer

Consider the problem of diffraction of a monochromatic field E^0, H^0 by a partially shielded layer. Denote by $D_1 = \mathrm{R}^3_+$ the halfspace, by $D_2 = U = \{\mathrm{x} : -1 < \mathrm{x}_3 < 0\}$, the layer, and by $\Omega \subset \mathrm{R}^2 = \{\mathrm{x}_3 = 0\}$, the domain bounded by the piece-wise smooth boundary $\Gamma = \partial\Omega$ formed by a finite number of simple C^∞-arcs that join at nonzero angles. We assume that the screens

$$\Sigma_1 = \{\mathrm{x} : \mathrm{x}_3 = 0,\ \mathrm{x} \in \mathrm{R}^2 \setminus \overline{\Omega}\}, \quad \Sigma_2 = \{\mathrm{x} : \mathrm{x}_3 = -1\}$$

are infinitely thin and perfectly conducting. The media D_1 and D_2 have the constant permittivity $\varepsilon = \varepsilon_j$ and permeability $\mu = \mu_j$ in D_j ($\omega > 0$) that satisfy the conditions listed in Section 4.3. The incident field E^0, H^0 satisfies the conditions specified by (4.50) and (4.51), which mean that the field sources are situated in R^3_+ and the field E^0, H^0 generated by these sources is located in $\overline{\mathrm{R}^3_+}$ shielded by the plane $x_3 = 0$. We will assume also that $E^0 \equiv 0$ and $H^0 \equiv 0$ in U.

Let us formulate the problem of diffraction by a partially shielded layer. It is necessary to determine the scattered field

$$E, H \in C^2(\mathrm{R}^3_+ \bigcup U) \bigcap_{\delta>0} C(\overline{\mathrm{R}^3_+} \setminus \Gamma_\delta) \bigcap_{\delta>0} C(\overline{U} \setminus \Gamma_\delta) \tag{4.72}$$

that satisfies the homogeneous Maxwell equations

$$\begin{aligned} Rot\, H &= -i\omega\varepsilon\, E, \\ Rot\, E &= i\omega\mu\, H, \quad \mathrm{x} \in \mathrm{R}^3_+ \bigcup U \end{aligned} \tag{4.73}$$

with $\varepsilon = \varepsilon_1$ and $\mu = \mu_1$ in R^3_+ and $\varepsilon = \varepsilon_2$ and $\mu = \mu_2$ in U, the boundary conditions

$$E_\tau|_\Sigma = 0, \quad \Sigma = \Sigma_1 \bigcup \Sigma_2 \tag{4.74}$$

for the tangential field components on Σ, the conjugation conditions on the interface

$$[E_\tau]_\Omega = 0, \tag{4.75}$$

$$[H_\tau]_\Omega = -[H^0_\tau]_\Omega, \tag{4.76}$$

and the condition of the finiteness of energy in every bounded volume

$$E, H \in L^2_{loc}(\mathrm{R}^3_+ \bigcup U). \tag{4.77}$$

Let us formulate now the conditions at infinity.
(i) $\mathrm{x} \in \mathrm{R}^3_+$:

$$E, H = o(r^{-1}), \quad r = |\mathrm{x}| \to \infty, \tag{4.78}$$

for $\Im\,\varepsilon_1 > 0$ or $\Im\,\mu_1 > 0$, and

$$E \times e_r + \left(\frac{\mu_1}{\varepsilon_1}\right)^{1/2} H = o(r^{-1}), \quad H = O(r^{-1}), \quad r \to \infty, \tag{4.79}$$

for $\Im\,\varepsilon_1 = 0$, $\Im\,\mu_1 = 0$, $\varepsilon_1 > 0$, and $\mu_1 > 0$, uniformly along all directions $\mathrm{x}/|\mathrm{x}|$.
(ii) $\mathrm{x} \in U$:

$$E, H = 0(\rho^{-1/2}), \quad \rho = |\mathrm{x}'| \to \infty \tag{4.80}$$

for $\Im\,\varepsilon_2 > 0$ or $\Im\,\mu_2 > 0$, uniformly along all directions x'/ρ and with respect to x_3. If $\Im\,\varepsilon_2 = 0$, $\Im\,\mu_2 = 0$, $\varepsilon_2 > 0$, and $\mu_2 > 0$, we will assume that the Fourier coefficients

$$\begin{aligned} u_n(\mathrm{x}') &= 2\int_{-1}^{0} u(\mathrm{x}) \cos \pi n x_3\, dx_3, \\ v_n(\mathrm{x}') &= 2\int_{-1}^{0} v(\mathrm{x}) \sin \pi n x_3\, dx_3, \quad n \geq 0, \end{aligned} \tag{4.81}$$

satisfy, for the components $u = E_\nu$ or $u = H_\tau$ and $v = H_\nu$ or $v = E_\tau$, the conditions

$$\frac{\partial}{\partial\rho}\begin{pmatrix}u_n\\ v_n\end{pmatrix} - ik_n\begin{pmatrix}u_n\\ v_n\end{pmatrix} = o(\rho^{-1/2}),\quad \begin{pmatrix}u_n\\ v_n\end{pmatrix} = O(\rho^{-1/2}),\quad \rho\to\infty \tag{4.82}$$

for $k_n^2 = k^2 - \pi^2 n^2 > 0$ (where $k_n > 0$ if $k > \pi n$ and $k_n < 0$ if $k < -\pi n$),

$$\begin{pmatrix}u_n\\ v_n\end{pmatrix} = O(1),\quad \rho\to\infty \tag{4.83}$$

for $k_n = 0$, and

$$\begin{pmatrix}u_n\\ v_n\end{pmatrix} = o(\rho^{-1/2}),\quad \begin{pmatrix}\partial u_n/\partial\rho\\ \partial v_n/\partial\rho\end{pmatrix} = o(\rho^{-1/2}),\quad \rho\to\infty \tag{4.84}$$

for $\Im k_n > 0$, uniformly along all directions x'/ρ and with respect to n. Here, $\nu = e_3 = (0,0,1)$ and $\nu\cdot\tau = 0$.

Relationships (4.82) constitute the Sommerfeld conditions for a bounded two-dimensional domain, and (4.83) is the condition at infinity for the two-dimensional Laplace equation. [4]. These conditions are imposed on a finite number of the Fourier coefficients; therefore, for them, a uniform behavior with respect to n is not required. But the requirement that estimates (4.84) are uniform with respect to n is essential and we will use it below.

The solution to problem (4.72)–(4.84) will be called the quasi-classical solution to the problem of diffraction by a partially shielded layer. For the total field in $\mathrm{R}^3_+ \cup U$, we have $E^{\mathrm{tot}} = E^0 + E$, $H^{\mathrm{tot}} = H^0 + H$.

Now we can prove the uniqueness of solution to problem (4.72)–(4.84).

Theorem 14 *For $\Im\varepsilon \geq 0$, $\Re\varepsilon > 0$, $\Im\mu \geq 0$, $\Re\mu > 0$, and $\omega > 0$, problem (4.72)–(4.84) has not more than one solution.*

□ The proof is based on the fact that homogeneous problem (4.72)–(4.84) has only a trivial solution.

Assume that $[H^0_\tau]_\Omega = 0$ and introduce the following convenient notations: $V_1 = \mathrm{R}^3_+ \cap B_R$, $S_1 = S^+_{\mathrm{R}}$, $V_2 = \{\mathrm{x} : |\mathrm{x}'| < R, \mathrm{x} \in U\}$ is a cylinder, $S_2 = \{\mathrm{x} : |\mathrm{x}'| = R,\ \mathrm{x} \in U\}$ is the lateral surface of the cylinder $V = V_1 \cup V_2$, and $S = S_1 \cup S_2$, where R is chosen so that $\overline{\Omega} \subset B_R$. Let us apply the Lorentz lemma to fields E and H in V. Taking into account the boundary conditions and the conjugation conditions, we obtain

$$\Re\int\limits_S (E\times\overline{H})n\,ds + \omega\int\limits_V (\Im\varepsilon\,|E|^2 + \Im\mu\,|H|^2)\,d\mathrm{x} = 0, \tag{4.85}$$

where n is the external normal to the surface. The possibility of applying the Lorentz lemma is justified in Section 2.1. Denote by

$$J_i = \Re\int\limits_{S_i} (E\times\overline{H})n\,ds,\quad i = 1,2$$

the useful scalar quantities. It is easy to see that when $\Im(\varepsilon_1 + \mu_1) > 0$, (4.78) yields $J_1 = o(1)$ as $R\to\infty$.

For real ε_1 and μ_1, we use (4.79) to obtain

$$J_1 = \left(\frac{\mu_1}{\varepsilon_1}\right)^{1/2} \int\limits_{S_1} |H|^2\, ds + o(1), \quad R \to \infty.$$

For the layer U, we use (4.80) to obtain $J_2 = o(1)$ as $R \to \infty$ when $\Im(\varepsilon_2 + \mu_2) > 0$.

In the case of real ε_2 and μ_2, we will assume that $\varepsilon_2 > 0$, $\mu_2 > 0$, $\Im k = 0$, and $k \neq 0$. Consider the fields $E = (E_\rho, E_\varphi, E_\nu)$ and $H = (H_\rho, H_\varphi, H_\nu)$ in the cylindrical coordinate system (ρ, φ, x_3). Substituting corresponding representations into the formula for J_2, we obtain

$$J_2 = R\Re \int\limits_0^{2\pi} d\varphi \int\limits_{-1}^{0} (E_\varphi \overline{H}_\nu - E_\nu \overline{H}_\varphi)\, dx_3.$$

Now, we substitute into the expression for J_2 the sine and cosine Fourier expansions, respectively, derived with respect to x_3 for the components E_ρ, E_φ, and H_ν and H_ρ, H_φ, and E_ν on the segment $[-1, 0]$. As a result, we obtain

$$J_2 = \frac{R}{2} \sum_{n=0}^{\infty} \frac{1}{1+\delta_{0n}} \Re \int\limits_0^{2\pi} \left(e_n^{(\varphi)} \overline{h_n^{(\nu)}} - e_n^{(\nu)} \overline{h_n^{(\varphi)}} \right) d\varphi,$$

where $e_n^{(\varphi)}$, $e_n^{(\nu)}$, $h_n^{(\varphi)}$, and $h_n^{(\nu)}$ are the Fourier coefficients ($h_0^{(\nu)} \equiv 0$ and $e_0^{(\varphi)} \equiv 0$). Note that here, the Fourier expansions are possible. Indeed, by virtue of Lemma 10, E and H are continuously differentiable in $\overline{U} \setminus V_2$.

Substituting these Fourier expansions into the Maxwell equations, we have

$$\begin{aligned} e_n^{(\varphi)} &= -\frac{1}{k_n^2} \left(\frac{\pi n}{R} \frac{\partial e_n^{(\nu)}}{\partial \varphi} + i\omega\mu_2 \frac{\partial h_n^{(\nu)}}{\partial \rho} \right), \\ h_n^{(\varphi)} &= \frac{1}{k_n^2} \left(\frac{\pi n}{R} \frac{\partial h_n^{(\nu)}}{\partial \varphi} + i\omega\varepsilon_2 \frac{\partial e_n^{(\nu)}}{\partial \rho} \right), \end{aligned}$$

where $k_n \neq 0$. Then, applying conditions (4.82) and (4.84), we obtain

$$\begin{aligned} J_2 &= \frac{R}{2} \sum_{n=0}^{\infty} \frac{1}{1+\delta_{0n}} \Re \left(-\frac{\pi n}{R k_n^2} \int\limits_0^{2\pi} \frac{\partial}{\partial \varphi} (e_n^{(\nu)} \overline{h_n^{(\nu)}})\, d\varphi \right. \\ &\quad \left. + \int\limits_0^{2\pi} \left(\frac{i\omega\varepsilon_2}{k_n^2} \overline{\frac{\partial e_n^{(\nu)}}{\partial \rho}} e_n^{(\nu)} - \frac{i\omega\mu_2}{k_n^2} \frac{\partial h_n^{(\nu)}}{\partial \rho} \overline{h_n^{(\nu)}} \right) d\varphi \right) \\ &= \frac{R}{2} \sum_{n:\pi^2 n^2 < k^2} \frac{1}{1+\delta_{0n}} \int\limits_0^{2\pi} \left(\frac{\omega\varepsilon_2}{k_n} \left| e_n^{(\nu)} \right|^2 + \frac{\omega\mu_2}{k_n} \left| h_n^{(\nu)} \right|^2 \right) d\varphi + o(1) \\ &\quad + \frac{R}{2} \sum_{n:\pi^2 n^2 > k^2} \Re \int\limits_0^{2\pi} \left(\frac{i\omega\varepsilon_2}{k_n^2} \overline{\frac{\partial e_n^{(\nu)}}{\partial \rho}} e_n^{(\nu)} - \frac{i\omega\mu_2}{k_n^2} \frac{\partial h_n^{(\nu)}}{\partial \rho} \overline{h_n^{(\nu)}} \right) d\varphi \\ &= \frac{R}{2} \sum_{n:\pi^2 n^2 < k^2} \frac{1}{1+\delta_{0n}} \frac{\omega}{k_n} \int\limits_0^{2\pi} \left(\varepsilon_2 \left| e_n^{(\nu)} \right|^2 + \mu_2 \left| h_n^{(\nu)} \right|^2 \right) d\varphi + o(1) \\ &\quad + \frac{R}{2}\, o(R^{-1}) \sum_{n:\pi^2 n^2 > k^2} \frac{1}{k_n^2} \\ &= \frac{R}{2} \sum_{n:\pi^2 n^2 < k^2} \frac{1}{1+\delta_{0n}} \frac{\omega}{k_n} \int\limits_0^{2\pi} \left(\varepsilon_2 \left| e_n^{(\nu)} \right|^2 + \mu_2 \left| h_n^{(\nu)} \right|^2 \right) d\varphi + o(1) \end{aligned}$$

for $R \to \infty$.

Substituting the expressions for J_i into (4.85) and using Lemma 9, we prove that $E \equiv 0$ and $H \equiv 0$ in R^3_+. Then, the conjugation condition (4.75) completes the formulation of the homogeneous Dirichlet and the homogeneous Neumann problems for E_τ and H_ν and for H_τ and E_ν, respectively, in the layer U for the Helmholtz equation with real parameter k^2 and conditions at infinity (4.80)–(4.84). As has been shown in Ref. [85], at $k^2 \neq \pi^2 n^2$, these problems have only the trivial solution.

If, for a certain n, $k^2 = \pi^2 n^2$, we perform a separate analysis of the term

$$J_2^0 = \frac{R}{2}\Re\int_0^{2\pi}\left(e_n^{(\varphi)}\overline{h_n^{(\nu)}} - e_n^{(\nu)}\overline{h_n^{(\varphi)}}\right)d\varphi, \quad k_n = 0.$$

From the Maxwell equations, we obtain

$$\begin{aligned} e_n^{(\nu)} &= -\frac{1}{i\omega\varepsilon_2}\left(\frac{1}{R}h_n^{(\varphi)} + \frac{\partial h_n^{(\varphi)}}{\partial\rho} - \frac{1}{R}\frac{\partial h_n^{(\rho)}}{\partial\varphi}\right), \\ h_n^{(\nu)} &= \frac{1}{i\omega\mu_2}\left(\frac{1}{R}e_n^{(\varphi)} + \frac{\partial e_n^{(\varphi)}}{\partial\rho} - \frac{1}{R}\frac{\partial e_n^{(\rho)}}{\partial\varphi}\right). \end{aligned}$$

Then,

$$J_2^0 = \frac{R}{2}\Re\int_0^{2\pi}\left(\frac{1}{i\omega\varepsilon_2 R}\left|h_n^{(\varphi)}\right|^2 - \frac{1}{i\omega\mu_2 R}\left|e_n^{(\varphi)}\right|^2\right)d\varphi + O(R^{-2}) = O(R^{-2}),$$

as $R \to \infty$, since the first term in the expression for J_2^0 vanishes identically and formula (4.83) yields (according to Ref. [4])

$$\frac{\partial e_n^{(\varphi)}}{\partial\rho}, \ \frac{1}{R}\frac{\partial e_n^{(\rho)}}{\partial\varphi}, \ \frac{\partial h_n^{(\varphi)}}{\partial\rho}, \ \frac{1}{R}\frac{\partial h_n^{(\rho)}}{\partial\varphi} = O(R^{-2}), \quad R \to \infty.$$

We see that at $k^2 = \pi^2 n^2$ and for a certain n, J_2 can be also represented as the sum of a positive term and a term that tends to infinity as $R \to \infty$, which again enables us to formulate the homogeneous Dirichlet and the homogeneous Neumann problems in the layer U for the components of E and H.

Using the results of Ref. [85], we can show that at $k^2 = \pi^2 n^2$, the Dirichlet and the Neumann problems in the layer U, have, by virtue of (4.83), only the standing-wave solutions. In particular, for the field H, we have $H_\tau = C\cos\pi n x_3$ and $H_\nu = 0$, where C does not depend on x. But the conjugation condition (4.76) yields $H_\tau|_\Omega = 0$; therefore, $C = 0$ and $H = 0$ in U, so that $E \equiv 0$ in U due to the second Maxwell equation (4.73).

Thus, in all cases, $E \equiv 0$ and $H \equiv 0$ in $R^3_+ \cup U$, which completes the proof of the theorem. □

We will look for a solution to problem (4.72)–(4.84) in the form

$$E = E_1, \ H = H_1 \quad \text{for} \quad x \in R^3_+, \tag{4.86}$$

$$E = E_2, \ H = H_2 \quad \text{for} \quad x \in U, \tag{4.87}$$

$$E_j = \frac{(-1)^{j+1}}{2\pi} Rot\,\tilde{A}_j u, \tag{4.88}$$

$$H_j = \frac{(-1)^{j+1}}{2\pi i\omega\mu_j}\left(Grad\, Div\, \tilde{A}_j u + k_j^2\, \tilde{A}_j u\right), \tag{4.89}$$

$$\tilde{A}_1 u = 2\pi \int_\Omega G_1^R(\mathrm{x}, y)\, u(y)\, dy = \int_\Omega \frac{e^{ik_1|\mathrm{x}-y|}}{|\mathrm{x}-y|}\, u(y)\, dy, \tag{4.90}$$

$$\tilde{A}_2 u = 2\pi \int_\Omega G_2^U(\mathrm{x}, y)\, u(y)\, dy = 0. \tag{4.91}$$

We will assume that the vector-function $u = (u^1, u^2)^T$ satisfies the conditions

$$u,\ div\, u \in C^1(\Omega), \quad u \in W(\overline{\Omega}). \tag{4.92}$$

Here, G_1^R and G_2^U are the Green's functions of (4.5) and (4.11)–(4.12), respectively. From formula (4.36), we obtain

$$G_2^U(\mathrm{x}, y) = G_2^R(\mathrm{x}, y) + V_2^U(\mathrm{x}, y), \quad V_2^U \in C^\infty(U_0 \times \overline{\Omega}),$$

$$\tilde{A}_2 u = \int_\Omega \frac{e^{ik_2|\mathrm{x}-y|}}{|\mathrm{x}-y|} u(y)\, dy + 2\pi \int_\Omega V_2^U(\mathrm{x}, y) u(y)\, dy, \quad \mathrm{x} \in U. \tag{4.93}$$

Since the Green's function G^U is not defined at $k_2 \in \Lambda(U)$, representation (4.91) is valid only when $k_2 \notin \Lambda(U)$. We will not construct modified representations for $k_2 \in \Lambda(U)$, limiting ourselves to the most important case $k_2 \notin \Lambda(U)$ (different methods for constructing modified representations are described in Refs. [27, 45, 89]).

Representations (4.86)–(4.91) and (4.93) supplied with conditions (4.92) are analyzed in Section 2.3. Obviously, an additional differentiable term in (4.93) does not lead to substantial changes. In the proof of the equality $Div\, \tilde{A}_2 u = \tilde{A}_2(div\, u)$, which is similar to that performed in Section 2.3, we take into account the symmetry of the Green's function: $G^U(\mathrm{x}, y) = G^U(y, \mathrm{x})$. The field H admits the representation

$$H_j = \frac{(-1)^{j+1}}{2\pi i\omega\mu_j}\left(Grad\, \tilde{A}_j\,(div\, u) + k_j^2\, \tilde{A}_j u\right), \tag{4.94}$$

which is similar to (4.89). The fields $E, H \in C^2(\mathrm{R}^3_+ \cup U)$ satisfy the Maxwell equations (4.73) and condition (4.77) of the finiteness of energy. The components H_τ and E_ν, defined by formulas (4.88) and (4.94), are continuous up to Ω from each side of the surface (except for the points on Γ). The differentiation and limiting transition can be performed under the integral sign, and the integral is understood in the sense of Cauchy. Limiting relationships (4.68) remain valid.

We use formulas (4.68) to extend the definitions of H_ν and E_τ to each side of Ω. According to the results of Section 2.3, one can perform the limiting transition $\mathrm{x} \to x_0 \in \mathrm{R}^2 \setminus \overline{\Omega}$, $\mathrm{x}_3 \to \pm 0$ in (4.88) and (4.89) under the integral sign. Therefore, fields E and H are continuous on each side of the plane $\mathrm{x}_3 = 0$ for $x_0 \in \mathrm{R}^2 \setminus \overline{\Omega}$. The same statement holds for the limiting transition $\mathrm{x}_3 \to -1$ and the continuity of E_2 and H_2 at the points of the plane $\mathrm{x}_3 = -1$. Boundary condition (4.74) for $x_0 \in \Sigma$ is satisfied because we have chosen the Neumann boundary conditions (4.2) and (4.7) for the Green's function. Thus, conditions (4.72)–(4.74), (4.75), and (4.77) are fulfilled, for every function u that satisfies (4.92), due to the choice of the solution in the form (4.86)–(4.91).

Now, we have to verify only the conditions at infinity. Conditions (4.78) and (4.79) in the halfspace are similar to the conditions at infinity considered in Section 4.3. The validity of (4.78) is obvious. Condition (4.79) follows from the Sommerfeld conditions (4.4), which are valid for the Green's function, and from the fact that, according to [23, 27], the Silver–Müller conditions (4.79) are equivalent to the Sommerfeld conditions. We see that, in the layer U, it is more convenient to represent, for $\Im k_2 > 0$, the Green's function G^U by means of formula (4.12). Then, conditions at infinity (4.80) are satisfied, because one can easily prove the simple estimates

$$O\left(\rho^{-1}e^{-\frac{\rho}{2}\Im k-|j|\Im k}\right), \quad j \in \mathrm{Z},$$

for the coefficients of series (4.12) and their derivatives $\partial/\partial \mathrm{x}_i$, which are uniform with respect to $\rho = |\mathrm{x}'|$ for $\rho \geq \rho_0$ and $y \in \overline{\Omega}$.

For $\Im \varepsilon_2 = 0$ and $\Im \mu_2 = 0$ ($\Im k_2 = 0$ and $k_2 \neq 0$ or $\Re k_2 = 0$ and $\Im k_2 > 0$) we use representation (4.11) for the Green's function G^U. From (4.11), we find

$$\begin{aligned}(\tilde{A}_2 u)_n(\mathrm{x}') &= 2\int_{-1}^{0} (\tilde{A}_2 u)(\mathrm{x}) \cos \pi n \mathrm{x}_3 d\mathrm{x}_3 \\ &= \pi i \int_{\Omega} H_0^{(1)}(k_n|\mathrm{x}'-y|)u(y)\,dy, \qquad (4.95)\end{aligned}$$

$$\begin{aligned}\left(\frac{\partial}{\partial \mathrm{x}_3}\tilde{A}_2 u\right)_n(\mathrm{x}') &= 2\int_{-1}^{0}\left(\frac{\partial}{\partial \mathrm{x}_3}\tilde{A}_2 u\right)(\mathrm{x}) \sin \pi n \mathrm{x}_3\,d\mathrm{x}_3 \\ &= -\pi^2 i n \int_{\Omega} H_0^{(1)}(k_n|\mathrm{x}'-y|)u(y)\,dy. \qquad (4.96)\end{aligned}$$

The Hankel function and its derivatives of the order $\alpha = (\alpha_1, \alpha_2)$ with respect to x_1 and x_2 satisfy the Sommerfeld conditions at $\Im k_n = 0$ and $k_n \neq 0$ [8, 8.451.3]:

$$\left(\frac{\partial}{\partial \rho} - ik_n\right)\partial_{\mathrm{x}'}^{\alpha} H_0^{(1)}(k_n|\mathrm{x}'-y|) = o(\rho^{-1/2}),$$

$$\partial_{\mathrm{x}'}^{\alpha} H_0^{(1)}(k_n|\mathrm{x}'-y|) = O(\rho^{-1/2}), \quad \rho = |\mathrm{x}'| \to \infty, \qquad (4.97)$$

uniformly with respect to $y \in \overline{\Omega}$. For $\Re k_n = 0$ and $\Im k_n > 0$, we have, according to [8, 8.451.6],

$$\partial_{\mathrm{x}'}^{\alpha} H_0^{(1)}(k_n|\mathrm{x}'-y|) = O\left(|k_n|^{|\alpha|-1}\rho^{-1}e^{-|k_n|\rho/2}\right), \qquad (4.98)$$

$$\rho \to \infty, \quad k_n = \pi i n + O(n^{-1}), \qquad (4.99)$$

uniformly with respect to $y \in \overline{\Omega}$ ($|\alpha| = \alpha_1 + \alpha_2$).

For the function

$$F_n = \partial_{\mathrm{x}'}^{\alpha} \int_{\Omega} H_0^{(1)}(k_n|\mathrm{x}'-y|)u(y)\,dy, \quad |\alpha| \leq 3,$$

we have the estimates

$$\begin{aligned}\left|\frac{\partial F_n}{\partial \rho} - ik_n F_n\right| &= \left|\left(\frac{\partial \Phi_n}{\partial \rho} - ik_n \Phi_n, u\right)\right| \\ &\leq \left\|\frac{\partial \Phi_n}{\partial \rho} - ik_n\Phi_n\right\|_{1/2} \|u\|_{-1/2} \\ &\leq \left\|\frac{\partial \Phi_n}{\partial \rho} - ik_n\Phi_n\right\|_{1} \|u\|_{1/2}, \\ |F_n| &\leq \|\Phi\|_{1/2}\|u\|_{-1/2} \leq \|\Phi\|_1 \|u\|_{-1/2},\end{aligned}$$

where $\Phi_n(\mathrm{x}', y) = \partial^{\alpha}_{\mathrm{x}'} H^{(1)}_0(k_n|\mathrm{x}' - y|)$. Combining these estimates with (4.97) and (4.99), we conclude that

$$\frac{\partial F_n}{\partial \rho} - ik_n F_n = o(\rho^{-1/2}), \quad F_n = O(\rho^{-1/2}), \quad \rho \to \infty,$$

for $\Im k_n = 0$ and $k_n \neq 0$, and

$$F_n, \frac{\partial F_n}{\partial \rho} = O\left(|k_n|^{|\alpha|}\rho^{-1}e^{-|k_n|\rho/2}\right), \quad \rho \to \infty,$$

for $\Re k_n = 0$ and $\Im k_n > 0$. Hence, (4.82) and (4.84) are satisfied, because, by virtue of representations (4.88), (4.94), and (4.91), the Fourier coefficients (4.81) of the E and H field components are expressed through functions (4.95) and (4.96) and the first derivatives of function (4.95) with respect to x_1 and x_2.

Thus, representation (4.86)–(4.91) of the fields guarantees the fulfillment of all conditions in diffraction problem (4.72)–(4.84), except for (4.76), for every function u that satisfies (4.92). The conjugation condition (4.76) yields the integrodifferential equation on Ω:

$$\sum_{j=1}^{2} \mu_j^{-1}\left(grad\, \tilde{A}_j\,(div\, u) + k_j^2\, \tilde{A}_j u\right) = f, \quad x \in \Omega, \quad u \in W, \tag{4.100}$$

$$\tilde{A}_1 u = A_1 u = \int_{\Omega} \frac{e^{ik_1|x-y|}}{|x-y|} u(y)\, dy, \tag{4.101}$$

$$\begin{aligned} \tilde{A}_2 u &= A_2 u + V_2 u \\ &= \int_{\Omega} \frac{e^{ik_1|x-y|}}{|x-y|} u(y)\, dy + 2\pi \int_{\Omega} V_2^U(x,y) u(y)\, dy, \end{aligned} \tag{4.102}$$

$$f(x) = -2\pi i\omega [H^0_\tau]_\Omega, \quad f \in C^\infty(\overline{\Omega}). \tag{4.103}$$

We have $div\, \tilde{A}_2 u = \tilde{A}_2(div\, u)$ and $div\, A_2 u = A_2(div\, u)$; therefore, $div\, V_2 u = V_2(div\, u)$, and we can rewrite (4.100) as

$$Su = \sum_{j=1}^{2} \mu_j^{-1}\left(grad\, A_j\,(div\, u) + k_j^2\, A_j u\right) + B_2 u = f, \quad x \in \Omega, \tag{4.104}$$

where

$$B_2 u = \int_{\Omega} b_2(x,y)\, u(y)\, dy,$$

and

$$b_2 = \frac{2\pi}{\mu_2} \begin{pmatrix} \left(\frac{\partial^2}{\partial \mathrm{x}_1^2} + k_2^2\right) V_2^U(\mathrm{x},y) & \frac{\partial^2}{\partial \mathrm{x}_1 \partial \mathrm{x}_2} V_2^U(\mathrm{x},y) \\ \frac{\partial^2}{\partial \mathrm{x}_1 \partial \mathrm{x}_2} V_2^U(\mathrm{x},y) & \left(\frac{\partial^2}{\partial \mathrm{x}_2^2} + k_2^2\right) V_2^U(\mathrm{x},y) \end{pmatrix} \in C^\infty(\overline{\Omega} \times \overline{\Omega}). \tag{4.105}$$

Equation (4.104) is similar to equation (4.41) with $b_1 \equiv 0$. For $\Im \varepsilon \geq 0$, $\Re \varepsilon > 0$, $\Im \mu \geq 0$, $\Re \mu > 0$, and $\omega > 0$, the conditions of Theorem 11 are fulfilled. Repeating the considerations of Section 4.3, we use Theorems 11 and 14 to prove the following theorem.

Theorem 15 *For $\Im\varepsilon \geq 0$, $\Re\varepsilon > 0$, $\Im\mu \geq 0$, $\Re\mu > 0$, $\omega > 0$, and $k_2 \notin \Lambda(U)$, equation (4.100) is uniquely solvable for arbitrary right-hand side $f \in W'$, in particular, for $f \in C^\infty(\overline{\Omega})$. Operator $S : W \to W'$ defined by formula (4.104) is continuously invertible. Problem (4.72)–(4.84) has the unique solution, which can be represented in the form (4.86)–(4.91), where u satisfies condition (4.100).*

Now, we can prove that the principle of limiting absorption is valid for the problem of diffraction by a partially shielded layer.

Statement 29 *Assume that $\Im\mu_j = 0$, $\Im k_j^0 = 0$, and $\mu_j > 0$ ($j = 1,2$) and $k_1^0 \neq 0$, $k_2^0 \notin \Lambda(U)$, and $\omega > 0$ are fixed. Then, if $f(k_1,k_2) \xrightarrow{W'} f(k_1^0,k_2^0)$ for $k_j \to k_j^0$ and $\Im k_j > 0$, we have $u(k_1,k_2) \xrightarrow{W} u(k_1^0,k_2^0)$, $E(k_1,k_2) \to E(k_1^0,k_2^0)$, and $H(k_1,k_2) \to H(k_1^0,k_2^0)$ in $L^2_{loc}(\mathrm{R}^3_+ \cup U)$ for $k_j \to k_j^0$ ($j = 1,2$), where u is the solution to equation (4.100), and E, H is the solution to problem (4.72)–(4.84) for the corresponding parameters ε, μ, and k.*

□ Let k_j and k_j^0 satisfy the conditions of the theorem. By virtue of Theorem 15, operator $S(k_1^0,k_2^0)$ is continuously invertible. The continuity of the operator-valued function $L = L(k_1,k_2)$ in (4.46) with respect to $k_1, k_2 \in \mathrm{C}$ is proved (in a more general situation) in Section 3.5. According to the results of [85] and representations (4.13), the continuous dependence of $B = B(k_2)$ in (4.46) follows from the continuity of function $V_2^U(\mathrm{x},y)$ in (4.93) and all its derivatives with respect to x_j and y_i on the parameter $k_2 \in \overline{\mathrm{C}}_+ \setminus \Lambda(U)$, $\mathrm{x}, y \in \overline{\Omega}$. We see that $S(k_1,k_2)$ satisfies the condition of Statement 27, which yields the required convergence $u(k_1,k_2) \xrightarrow{W} u(k_1^0,k_2^0)$ as $k_j \to k_j^0$.

The continuous dependence of the operator-valued function

$$\tilde{A}_1(k_1) : \tilde{H}^{-1/2}(\overline{\Omega}) \to H^1_{loc}(\mathrm{R}^3_+)$$

on $k_1 \in \mathrm{C}$ and the continuity of the first term in (4.93) on $k_2 \in \mathrm{C}$ are proved in Section 3.5. In [85], it has been shown that $V_2^U(\mathrm{x},y)$ in (4.93) and all its derivatives with respect to x_j and y_i are continuous functions of $k_2 \in \overline{\mathrm{C}}_+ \setminus \Lambda(U)$ uniformly if x, y belong to a bounded subset of $\overline{U}$ and k_2, to a compact set in $\overline{\mathrm{C}}_+ \setminus \Lambda(U)$. Repeating the corresponding part of the proof of Theorem 10, we find that the operator-valued function

$$\tilde{A}_2(k_2) : \tilde{H}^{-1/2}(\overline{\Omega}) \to H^1_{loc}(\mathrm{R}^3_-)$$

is continuous with respect to k_2 in $\overline{\mathrm{C}}_+ \setminus \Lambda(U)$. Then, representations (4.88) and (4.94) yield the continuity of fields E and H with respect to $k_1 \in \mathrm{C}$ and $k_2 \in \overline{\mathrm{C}}_+ \setminus \Lambda(U)$ in the norm $L^2_{loc}(\mathrm{R}^3_+ \cup U)$. □

4.5 Diffraction by an aperture in a semi-infinite waveguide

Consider the problem of diffraction of a monochromatic field E^0, H^0 by an aperture in a waveguide coupled with the half-space. Denote by $D_1 = \mathrm{R}^3_+$ the halfspace and by $D_2 = T^-$, the semi-infinite rectangular cylinder (according to the definition of Section 4.3, this cylinder may be called the waveguide domain). Let $\Omega \subset \mathrm{R}^2 = \{\mathrm{x}_3 = 0\}$ be the bounded domain such that $\overline{\Omega} \subset \Pi = \{\mathrm{x} : |\mathrm{x}_1| < a/2, |\mathrm{x}_2| < b/2, \mathrm{x}_3 = 0\}$. We assume that its boundary $\Gamma = \partial\Omega$ satisfies the conditions formulated in Section 4.3 and the screens $\Sigma_1 = \left\{\mathrm{x} : \mathrm{x}_3 = 0, \mathrm{x} \in \mathrm{R}^2 \setminus \overline{\Omega}\right\}$ and $\Sigma_2 = \partial T^- \setminus \overline{\Omega}$ ($\Sigma = \Sigma_1 \cup \Sigma_2$) are infinitely thin and perfectly conducting. Note that $\partial\Omega \bigcap \partial\Pi = \emptyset$. The media in D_1 and D_2 have the constant permittivity $\varepsilon = \varepsilon_j$ and permeability $\mu = \mu_j$ that satisfy the conditions listed in Section 4.3.

Assume that the incident field E^0, H^0 is a solution to the Maxwell equations in T^- with the boundary condition

$$E_\tau^0\Big|_{\partial T^-} = 0, \tag{4.106}$$

which is imposed on smooth parts of surface ∂T^-. The incident field is generated by sources situated outside $\overline{\Omega}$, so that

$$H_\tau^0\Big|_\Omega \in C^\infty(\overline{\Omega}). \tag{4.107}$$

Usually, E^0, H^0 is a certain waveguide mode that comes from infinity [2]. The field E^0, H^0 in R^3_+ vanishes identically. One could also consider the problem when the field sources are placed in the halfspace R^3_+; however, this situation is not typical for practical applications. Anyhow, it is sufficient to apply condition (4.107), since we do not use any other information about the incident field.

The problem of diffraction by an aperture in a semi-infinite rectangular waveguide is reduced to determination of the scattered electromagnetic field

$$E, H \in C^2(R^3_+ \bigcap T^-) \bigcap_{\delta>0} C(\overline{R^3_+} \setminus \Gamma_\delta) \bigcap_{\delta>0} C(\overline{T^-} \setminus \Gamma_\delta), \tag{4.108}$$

that satisfies the Maxwell equations

$$\begin{aligned} Rot H &= -i\omega\varepsilon\, E, \\ Rot E &= i\omega\mu\, H, \quad x \in R^3_+ \bigcup T^-, \end{aligned} \tag{4.109}$$

where $\varepsilon = \varepsilon_1$ and $\mu = \mu_1$ in R^3_+ and $\varepsilon = \varepsilon_2$ and $\mu = \mu_2$ in T^-, the boundary conditions

$$E_\tau|_\Sigma = 0 \tag{4.110}$$

for the tangential field components on Σ (which are imposed on smooth parts of surface Σ), the conjugation conditions on the interface

$$[E_\tau]_\Omega = 0, \tag{4.111}$$

$$[H_\tau]_\Omega = -[H_\tau^0]_\Omega, \tag{4.112}$$

and the condition of the finiteness of energy in every bounded volume

$$E, H \in L^2_{loc}(R^3_+ \bigcup T^-). \tag{4.113}$$

Let us formulate the conditions at infinity.
(i) $x \in R^3_+$:

$$E, H = o(r^{-1}), \quad r = |x| \to \infty, \tag{4.114}$$

for $\Im\,\varepsilon_1 > 0$ or $\Im\,\mu_1 > 0$, and

$$E \times e_r + \left(\frac{\mu_1}{\varepsilon_1}\right)^{1/2} H = o(r^{-1}), \quad H = O(r^{-1}), \quad r \to \infty, \tag{4.115}$$

for $\Im\,\varepsilon_1 = 0$, $\Im\,\mu_1 = 0$, $\varepsilon_1 > 0$, and $\mu_1 > 0$, uniformly along all directions $x/|x|$.

(ii) $x \in T^-$:

$$E, H = o(1), \quad r \to \infty, \tag{4.116}$$

for $\Im \varepsilon_2 > 0$ or $\Im \mu_2 > 0$, and the fields E and H admit for $x_3 < 0$ the representations

$$\begin{aligned}\begin{pmatrix} E \\ H \end{pmatrix} &= \sum_p R_p^{(1)} e^{-i\gamma_p^{(1)} x_3} \begin{pmatrix} \lambda_p^{(1)} \Pi_p e_3 - i\gamma_p^{(1)} \nabla_2 \Pi_p \\ -i\omega\varepsilon_2(\nabla_2 \Pi_p) \times e_3 \end{pmatrix} \\ &+ \sum_p R_p^{(2)} e^{-i\gamma_p^{(2)} x_3} \begin{pmatrix} i\omega\mu_2(\nabla \Psi_p) \times e_3 \\ \lambda_p^{(2)} \Psi_p e_3 - i\gamma_p^{(2)} \nabla_2 \Psi_p \end{pmatrix},\end{aligned} \tag{4.117}$$

for $\Im \varepsilon_2 = 0$, $\Im \mu_2 = 0$, $\varepsilon_2 > 0$, and $\mu_2 > 0$. Here, $\gamma_p^{(j)} = \sqrt{k^2 - \lambda_p^{(j)}}$, $\Im \gamma_p^{(j)} > 0$ or $\Im \gamma_p^{(j)} = 0$, $k\gamma_p^{(j)} \geq 0$; $\lambda_p^{(1)}$, $\Pi_p(x_1, x_2)$ and $\lambda_p^{(2)}$, $\Psi(x_1, x_2)$ are the complete systems of eigenvalues and orthogonal and normalized in $L_2(\Pi)$ eigenfunctions of the two-dimensional Laplace operator $-\Delta$ in the rectangle Π with the Dirichlet and the Neumann conditions, respectively; and $\nabla_2 \equiv e_1\, \partial/\partial x_1 + e_2\, \partial/\partial x_2$. The estimates

$$R_p^{(1)}, R_p^{(2)} = O(p^m), \quad p \to \infty, \tag{4.118}$$

are valid, for a certain $m \in \mathbb{N}$, for the coefficients of series (4.117).

From the physical viewpoint, conditions (4.117) mean that the scattered field is a superposition of normal waves that radiate off the aperture [23]. Conditions (4.118) provide an exponential convergence of series (4.117), as well as the possibility of their termwise differentiation with respect to x_j arbitrary number of times.

We will call the solution to problem (4.108)–(4.118) the quasi-classical solution to the diffraction problem in a semi-infinite waveguide. For the total field in $\mathbb{R}^3_+ \bigcup T^-$, we have $E^{\text{tot}} = E^0 + E$ and $H^{\text{tot}} = H^0 + H$.

Let us prove the uniqueness of solution to problem (4.108)–(4.118).

Theorem 16 *For $\Im \varepsilon \geq 0$, $\Re \varepsilon > 0$, $\Im \mu \geq 0$, $\Re \mu > 0$, and $\omega > 0$, problem (4.108)–(4.118) has not more than one solution.*

□ In order to prove the theorem, it is sufficient to show that homogeneous problem (4.108)–(4.118) has only a trivial solution

Assume that $[H_\tau^0]_\Omega = 0$ and introduce the notations $V_1 = \mathbb{R}^3_+ \bigcap B_R$, $S_1 = S_R^+$, $V_2 = \{x : |x_3| < R,\ x \in T^-\}$, $S_2 = \{x : |x_3| = R,\ x \in T^-\}$, $V = V_1 \bigcup V_2$, and $S = S_1 \bigcup S_2$, where R is chosen so that $\overline{\Omega} \subset B_R$. Applying the Lorentz lemma to fields E and H in V (which is justified in Section 2.1) and taking into account the boundary conditions and the conjugation conditions, we obtain

$$\Re \int_S (E \times \overline{H}) n\, ds + \omega \int_V (\Im \varepsilon\, |E|^2 + \Im \mu\, |H|^2)\, dx = 0, \tag{4.119}$$

where $n = -e_3$ is the external normal to the surface. Note that when the Lorentz lemma is applied, the fact that ∂V_2 is not a smooth but only a piecewise-smooth surface does not cause any difficulties. Indeed, according to conditions (4.108), fields E and H are continuous in the vicinity of edges, because a certain space (a positive distance) separates the line of singularities $\partial\Omega$ and $\partial\Pi$. Again, we denote by

$$J_i = \Re \int_{S_i} (E \times \overline{H}) n\, ds, \quad i = 1, 2,$$

a useful scalar quantity. It is easy to see that when $\Im(\varepsilon_1+\mu_1) > 0$, (4.114) yields $J_1 = o(1)$ as $R \to \infty$.

For real ε_1 and μ_1, we obtain, by virtue of (4.115), the estimate

$$J_1 = \left(\frac{\mu_1}{\varepsilon_1}\right)^{1/2} \int\limits_S |H|^2\, ds + o(1), \quad R \to \infty.$$

For a semi-infinite cylinder, from (4.116), we have $J_2 = o(1)$ as $R \to \infty$ for $\Im(\varepsilon_2 + \mu_2) > 0$.

For real $\varepsilon_2 > 0$ and $\mu_2 > 0$, we perform simple transformations to obtain

$$J_2 = \omega\varepsilon_2 \sum_{p:\gamma_p^{(1)}>0} \left|R_p^{(1)}\right|^2 \gamma_p^{(1)}\lambda_p^{(1)} + \omega\mu_2 \sum_{p:\gamma_p^{(2)}>0} \left|R_p^{(2)}\right|^2 \gamma_p^{(2)}\lambda_p^{(2)} \ (> 0).$$

Note that these sums are finite and do not depend on x. Substituting the expressions for J_i into (4.119) and taking into account Lemmas 9 and 10, we find that $E \equiv 0$ and $H \equiv 0$ in R^3_+. Then, the conjugation conditions (4.111) and (4.112) yield $E_\tau|_\Omega = 0$ and $H_\tau|_\Omega = 0$. From the condition $E_\tau|_\Omega = 0$ and the Maxwell equations, it follows that $H_\nu|_\Omega = 0$ and $\partial H_\tau/\partial \mathrm{x}_3|_\Omega = 0$. The equation $Div H = 0$ yields $\partial H_\nu/\partial \mathrm{x}_3|_\Omega = 0$. Thus, all components of the field H satisfies the homogeneous Helmholtz equation in T^- and the homogeneous Dirichlet and the homogeneous Neumann conditions on Ω (which constitute the homogeneous Cauchy data). Finally, since H is analytical in T^-, we have, according to [34], that $H \equiv 0$ in T^- and, consequently, $E \equiv 0$ in T^- due to the first Maxwell equation (4.109). □

We will look for a solution to problem (4.108)–(4.118) in the form

$$E = E_1, \ H = H_1 \text{ for } \mathrm{x} \in \mathrm{R}^3_+, \tag{4.120}$$

$$E = E_2, \ H = H_2 \text{ for } \mathrm{x} \in T_-, \tag{4.121}$$

$$E_j = \frac{(-1)^{j+1}}{2\pi} Rot\, \tilde{A}_j u, \tag{4.122}$$

$$H_j = \frac{(-1)^{j+1}}{2\pi i\omega\mu_j}(Grad\, Div\, \tilde{A}_j u + k^2\, \tilde{A}_j u), \tag{4.123}$$

$$\tilde{A}_1 u = 2\pi \int\limits_\Omega G_1^R(\mathrm{x}, y)\, u(y)\, dy = \int\limits_\Omega \frac{e^{ik_1|\mathrm{x}-y|}}{|\mathrm{x}-y|} u(y)\, dy, \tag{4.124}$$

$$\tilde{A}_2 u = 2\pi \int\limits_\Omega \widehat{G}_2^T(\mathrm{x}, y) u(y)\, dy, \tag{4.125}$$

where $\widehat{G}_2^T = diag(G_a^T, G_b^T)$ is the matrix 2×2 Green's function, constructed from the scalar Green's functions (4.23) and (4.24). The vector-function $u = (u^1, u^2)^T$ satisfies the conditions

$$u,\ div\, u \in C^1(\Omega), \quad u \in W(\overline{\Omega}). \tag{4.126}$$

Using formulas (4.37) and (4.38), we obtain

$$\widehat{G}_2^T(\mathrm{x}, y) = G_2^R(\mathrm{x}, y)\hat{I} + \hat{V}_2^T(\mathrm{x}, y), \hat{V}_2^T \in C^\infty(\overline{T^-} \times \overline{\Omega}), \tag{4.127}$$

$$\tilde{A}_2 u = \int\limits_\Omega \frac{e^{ik_2|\mathrm{x}-y|}}{|\mathrm{x}-y|} u(y)\, dy + 2\pi \int\limits_\Omega \hat{V}_2(\mathrm{x}, y) u(y)\, dy, \tag{4.128}$$

where $\widehat{V_2^T} = diag\,(V_a^T, V_b^T)$ and $\hat{I}$ is the unit 2×2 matrix.

Since the Green's functions $G^T_{a,b}$ are not defined for $k_2 \in \Lambda_{a,b}(T)$, representation (4.125) can be applied only for $k_2 \notin \Lambda_a(T) \bigcup \Lambda_b(T)$. Here, we will not consider modified representations of the Green's functions for $k_2 \in \Lambda_a(T) \bigcup \Lambda_b(T)$.

Analysis of representations (4.120)–(4.125) and (4.127)–(4.128) with conditions (4.126) is similar to that performed in Section 2.3. We will indicate only the differences. Now, we consider the general case, when $div\,\tilde{A}_2 u \neq \tilde{A}_2(div\,u)$. This formula can be rewritten as

$$div\,\tilde{A}_2 u = \tilde{A}_2'(div\,u),$$

where

$$\tilde{A}_2'(div\,u) = 2\pi \int\limits_{\Omega} G^T(\mathrm{x}, y')div\,u(y')\,dy', \quad \mathrm{x} \in T^-,$$

and

$$G^T(\mathrm{x}, y') = G^T(\mathrm{x}, y)\Big|_{y_3=0}$$

is the trace of the Green's function $G^T(\mathrm{x}, y)$ of the Neumann problem for the Helmholtz equation (with the Neumann conditions imposed on the boundary ∂T^-):

$$G^T(\mathrm{x}, y) = \frac{2i}{ab}\sum_{n=0}^{\infty}\sum_{m=0}^{\infty} \frac{\Psi_{nm}(\mathrm{x}')\Psi_{nm}(y')}{(1+\delta_{0n})(1+\delta_{0m})\gamma_{nm}} \left(e^{i\gamma_{nm}|\mathrm{x}_3-y_3|} + e^{i\gamma_{nm}|\mathrm{x}_3+y_3|}\right),$$

$$\Psi_{nm}(z') = \cos\frac{\pi n}{a}\left(z_1 + \frac{a}{2}\right)\cos\frac{\pi m}{b}\left(z_2 + \frac{b}{2}\right).$$

For $u \in C_0^\infty(\Omega)$, this formula can be verified by elementary methods (indeed, the series converge exponentially as $\mathrm{x}_3 \neq 0$). Obviously, the Green's function has the same singularities as function $G^T_{a,b}$; therefore, this representation can be extended to elements $u \in W$ by continuity (see Section 2.3). Then, formula (4.123) takes the form

$$H_1 = \frac{1}{2\pi i\omega\mu_1}\left(Grad\,\tilde{A}_1(div\,u) + k_1^2\,\tilde{A}_1 u\right), \quad \mathrm{x} \in \mathrm{R}^3_+, \tag{4.129}$$

$$H_2 = \frac{-1}{2\pi i\omega\mu_2}\left(Grad\,\tilde{A}_2'(div\,u) + k_2^2\,\tilde{A}_2 u\right), \quad \mathrm{x} \in T^-. \tag{4.130}$$

This representation is equivalent to (4.123). Here,

$$\left(\tilde{A}_2' u\right)(\mathrm{x}) = \int\limits_{\Omega} \frac{e^{ik_2|\mathrm{x}-y|}}{|\mathrm{x}-y|}u(y)\,dy + \int\limits_{\Omega} V^T(\mathrm{x}, y)u(y)\,dy, \tag{4.131}$$

and $V^T \in C^\infty(\overline{T^-} \times \overline{\Omega})$. Formulas (4.129)–(4.131) are more convenient than (4.123) because they contain only one external operation of differentiation. We will use this property below, when the principle of limiting absorption will be derived.

When $\mathrm{x}_3 \neq 0$, we can differentiate the representations for E and H under the integral sign. The fields $E, H \in C^2\left(\mathrm{R}^3_+ \bigcup T^-\right)$ satisfy the Maxwell equations (4.109) and conditions (4.113). The components H_τ and E_ν ($\nu = e_3$) defined by formula (4.122) and (4.129)–(4.131) are continuous up to Ω from each side of the surface (except for the points on Γ). The differentiation and limiting transition can be performed under the integral sign, and the integral is understood in the sense of Cauchy. Limiting relationships (4.68) remain valid, because the Green's functions $G^T_{a,b}$ satisfy, as well as G^R, the Neumann boundary condition for $x_3 = 0$ at the points of Ω and have the same singularities in (4.127)–(4.131) as the Green's function in (4.65) and (4.67). From (4.68), it follows that the conjugation condition (4.111) is valid.

We use formulas (4.68) to define H_ν and E_τ on each side of Ω. The limiting transitions in formulas (4.122), (4.129), and (4.131) for $x \to x_0 \in R^2 \setminus \overline{\Omega}$ and $x_3 \to +0$ and $x \to x_0 \in \overline{\Pi} \setminus \overline{\Omega}$ and $x_3 \to -0$, can be performed under the integral sign, because, by virtue of Lemma 8, $V^T_{a,b} \in C^\infty(\overline{T^-} \times \overline{\Omega})$, and fields E and H are continuous at these points on the corresponding side of the plane $x_3 = 0$.

The possibility of the termwise differentiation of series (4.23) and (4.24) with respect to x_j, y_1, and y_2 at $y_3 = 0$ and $x_3 \neq 0$ arbitrary number of times and their exponential convergence enable us to conclude that E and H are continuous in $\overline{T^-} \setminus \{x_3 = 0\}$. We can also verify the fulfillment of boundary condition (4.110) on the lateral surface $\Sigma \setminus \{x_3 = 0\}$. Note that we have chosen mixed boundary conditions (4.16) and (4.20) for the Green's functions so that the field E represented by formulas (4.122), (4.124), and (4.125) satisfy condition (4.110).

Now, we can prove the validity of conditions at infinity. Conditions (4.114) and (4.115) in the halfspace, which are considered in Sections 4.3 and 4.4, coincide with (4.58) and (4.59) or with (4.78) and (4.79). Condition (4.116) in a semi-infinite cylinder is fulfilled because we have $\gamma_p^{(a)} \sim i\sqrt{\lambda_p^{(a)}}$ and $\gamma_p^{(b)} \sim i\sqrt{\lambda_p^{(b)}}$ as $p \to \infty$; these estimate provide, for $y_3 = 0$ and $x_3 \neq 0$, the uniform (with respect to x' and y') convergence of the series in (4.17) and (4.21) and the uniform convergence of their derivatives of arbitrary order with respect to x_j and y_i. Obviously, these series tend to zero as $|x_3| \to \infty$.

Let us consider condition (4.117). Using explicit forms of the Green's functions given by formulas (4.23) and (4.24), we derive representation (4.117) with the coefficients

$$R_p^{(1)} = R_{nm}^{(1)} = -\frac{1}{\lambda_{nm}}\left(\frac{\pi m}{b} a_{nm}^{(1)} - \frac{\pi n}{a} a_{nm}^{(2)}\right), \tag{4.132}$$

$$R_p^{(2)} = R_{nm}^{(2)} = \frac{\gamma_{nm}}{\omega \mu_2 \lambda_{nm}}\left(\frac{\pi n}{a} a_{nm}^{(1)} + \frac{\pi m}{b} a_{nm}^{(2)}\right), \tag{4.133}$$

where

$$a_{nm}^{(j)} = \frac{2i}{\gamma_{nm}\sqrt{ab}} \int_\Omega f_{nm}^j(y) u^j(y)\, dy; \quad u = (u^1, u^2)^T \in W. \tag{4.134}$$

Note that it is sufficient to verify the equality only for longitudinal components E_ν and H_ν in (4.116) and (4.117) and in representations (4.122), (4.123), (4.125), (4.23), and (4.24), since the field E, H is completely defined in T^- by E_ν and H_ν and is expressed as a superposition of normal waves [43, 44]. For coefficients (4.134), we obtain the estimates

$$\begin{aligned}
|a_{nm}^{(j)}| &= \frac{2}{|\gamma_{nm}|\sqrt{ab}} \left| \int_\Omega f_{nm}^j(y) u^j(y)\, dy \right| \\
&\leq \frac{2}{|\gamma_{nm}|\sqrt{ab}} \|f_{nm}^j\|_{H^{1/2}(\Omega)} \|u\|_{\tilde{H}^{-1/2}(\overline{\Omega})} \\
&\leq \frac{2}{|\gamma_{nm}|\sqrt{ab}} \|f_{nm}^j\|_{H^1(\Omega)} \|u\|_W \\
&= \frac{(1+\lambda_{nm})^{1/2}}{|\gamma_{nm}|} \|u\|_W \leq C\|u\|_W,
\end{aligned}$$

which are uniform with respect to n and m. Therefore, estimate (4.118) is also valid.

Thus, for the problem of diffraction in a semi-infinite cylinder, all conditions (4.108)–(4.118) are fulfilled, except for (4.112), due to the choice of representations (4.120)–(4.125) for fields E and H for arbitrary u that satisfy (4.126). The conjugation condition (4.112) yields the integrodifferential equation on Ω

$$Su = \sum_{j=1}^{2} \mu_j^{-1} \left(grad\, A_j(div\, u) + k_j^2\, A_j u \right) + B_2 u = f, \quad x \in \Omega, \tag{4.135}$$

$$A_j u = \int_{\Omega} \frac{e^{ik_j|x-y|}}{|x-y|} u(y)\, dy, \quad j = 1, 2, \tag{4.136}$$

$$B_2 u = \frac{2\pi}{\mu_2} \int_{\Omega} (grad\, div + k_2^2)_x \left(\hat{V}_2^T(x,y) u(y) \right) dy, \tag{4.137}$$

$$f(x) = -2\pi i\omega\, [H_\tau^0]_\Omega, \quad f \in C^\infty(\overline{\Omega}). \tag{4.138}$$

Equation (4.135) is similar to (4.41) with $B_1 = 0$, since the 2×2 matrix

$$b_2(x,y) = \frac{2\pi}{\mu_2} \begin{pmatrix} \left(\frac{\partial^2}{\partial x_1^2} + k_2^2 \right) V_a^T & \frac{\partial^2}{\partial x_1 \partial x_2} V_b^T \\ \frac{\partial^2}{\partial x_1 \partial x_2} V_a^T & \left(\frac{\partial^2}{\partial x_2^2} + k_2^2 \right) V_b^T \end{pmatrix} \in C^\infty(\overline{\Omega} \times \overline{\Omega})$$

is infinitely smooth.

Repeating the considerations of Section 4.3, we use Theorems 11 and 14 to prove the following theorem.

Theorem 17 *For $\Im \varepsilon \geq 0$, $\Re \varepsilon > 0$, $\Im \mu \geq 0$, $\Re \mu > 0$, $\omega > 0$, and $k_2 \notin \Lambda(T)$, equation (4.135) is uniquely solvable for arbitrary right-hand side $f \in W'$, in particular, for $f \in C^\infty(\overline{\Omega})$. Operator $S : W \to W'$ defined by formula (4.135) is continuously invertible. Problem (4.108)–(4.118) has the unique solution u, which can be represented in the form (4.120)–(4.125), where u satisfies condition (4.126) and equation (4.135).*

Let us prove that the principle of limiting absorption is valid for the problem of diffraction in a semi-infinite cylinder.

Statement 30 *Assume that $\Im \mu_j = 0$, $\Im k_j^0 = 0$, and $\mu_j > 0$ $(j = 1, 2)$ and $k_1^0 \neq 0$, $k_2^0 \notin \Lambda(T)$, and $\omega > 0$ are fixed. If $f(k_1, k_2) \xrightarrow{W'} f(k_1^0, k_2^0)$ for $k_j \to k_j^0$, $\Im k_j > 0$, then $u(k_1, k_2) \xrightarrow{W} u(k_1^0, k_2^0)$, $E(k_1, k_2) \to E(k_1^0, k_2^0)$, and $H(k_1, k_2) \to H(k_1^0, k_2^0)$ in $L^2_{loc}(\mathrm{R}^3_+ \cup T^-)$ for $k_j \to k_j^0$, $j = 1, 2$. u is the solution to equation (4.135) and E, H is the solution to problem (4.108)–(4.118) for the corresponding parameters ε, μ, and k.*

□ The proof of this statement is similar to that of Statement 29. In fact, according to representations (4.122) and (4.129)–(4.131), it is sufficient to check that the operator-valued function

$$\tilde{A}_2(k_2) : \tilde{H}^{-1/2}(\overline{\Omega}) \to H^1_{loc}(\mathrm{R}^3)$$

is continuous with respect to $k_2 \in \overline{\mathrm{C}}_+ \setminus (\Lambda_a(T) \cup \Lambda_b(T))$ and the operator-valued function

$$\tilde{A}_2'(k_2) : \tilde{H}^{-1/2}(\overline{\Omega}) \to H^1_{loc}(\mathrm{R}^3)$$

is continuous with respect to $k_2 \in \overline{\mathrm{C}}_+ \setminus (\Lambda_a(T) \cup \Lambda_b(T) \cup 0)$. The required continuity of $\tilde{A}_2$ and $\tilde{A}_2'$ follows from the estimate

$$\left\| \Phi(x, y; k_2, k_2^0) \right\|_{H^1_{loc}(\mathrm{R}^3) \times H^{1/2}(\Omega)} \leq C |k_2 - k_2^0|, \tag{4.139}$$

which is uniform with respect to k_2 for the function

$$\Phi(\mathrm{x}, y; k_2, k_2^0) = G(\mathrm{x}, y; k_2) - G(\mathrm{x}, y; k_2^0),$$

where G is any of the functions G_a^T, G_b^T, or G^T.

Let us estimate the norm on the left-hand side of (4.139). Φ, considered as a function of y, is defined in a wider domain Π, and

$$\|\Phi\|_{H^1_{loc}(\mathrm{R}^3)\times H^{1/2}(\Omega)} \leq \|\Phi\|_{H^1_{loc}(\mathrm{R}^3)\times H^{1/2}(\Pi)} . \tag{4.140}$$

But Φ can be represented as the Fourier series over the orthogonal and normalized trigonometric basis in the rectangle Π, so that we can determine the quantity on the left-hand side of (4.140). Set

$$D_0 = \left\{\mathrm{x} : -R < \mathrm{x}_3 < 0, \mathrm{x} \in T^-\right\}.$$

Then, according to [7, 22],

$$\|\Phi\|_{H^1(D_0)\times H^{1/2}(\Pi)} = \int_{-R}^{0} \sum_{n,m} (1+\lambda_{nm})^{1/2} \left((1+\lambda_{nm}) \left| \frac{e^{i\gamma_{nm}|\mathrm{x}_3|}}{\gamma_{nm}} - \frac{e^{i\gamma^0_{nm}|\mathrm{x}_3|}}{\gamma^0_{nm}} \right|^2 + \left| e^{i\gamma_{nm}|\mathrm{x}_3|} - e^{i\gamma^0_{nm}|\mathrm{x}_3|} \right|^2 \right) d\mathrm{x}_3, \tag{4.141}$$

where $\gamma_{nm} = \sqrt{k_2^2 - \lambda_{nm}}$ and $\gamma^0_{nm} = \sqrt{(k_2^0)^2 - \lambda_{nm}}$. Using a simple inequality

$$\left| t^m e^{-(\alpha+i\beta)t} \right| \leq \frac{m^m}{\alpha^m e},$$

where $m \geq 0$ is an integer, $\alpha > 0$, $\beta \in \mathrm{R}^1$, and $t \in [0, \infty)$, we obtain

$$e^{i\gamma_{nm}|\mathrm{x}_3|} - e^{i\gamma^0_{nm}|\mathrm{x}_3|} = (k_2 - k_2^0) e^{i\gamma^0_{nm}|\mathrm{x}_3|} \frac{i|\mathrm{x}_3|(k_2 + k_2^0)}{2\gamma^0_{nm}} + (k_2 - k_2^0)^2 O(|\gamma^0_{nm}|^{-4}) = (k_2 - k_2^0)\, O(|\gamma^0_{nm}|^{-2}),$$

and, consequently,

$$\frac{e^{i\gamma_{nm}|\mathrm{x}_3|}}{\gamma_{nm}} - \frac{e^{i\gamma^0_{nm}|\mathrm{x}_3|}}{\gamma^0_{nm}} = (k_2 - k_2^0)\, O(|\gamma^0_{nm}|^{-3}),$$

uniformly with respect to k_2. Taking into account the asymptotic relationship $|\gamma^0_{nm}|^{-1} = O((n^2+m^2)^{-1/2})$ as $n, m \to \infty$, we conclude that series (4.141) converges and estimate (4.139) is valid. □

Bibliography

1. V.M. Babich, V.S. Buldyrev. *Asymptotic methods in the problems of short wave diffraction.* Nauka, Moscow, (1972) (in Russian).
2. L.A. Weinschtien. *Electromagnetic waves.* Radio i Svyaz', Moscow, (1988) (in Russian).
3. E.N. Vasil'ev. *Excitation of bodies of rotation.* Radio i Svyaz', Moscow, (1987) (in Russian).
4. V.S. Vladimirov. *Equations of mathematical physics.* Nauka, Moscow, (1981) (in Russian).
5. V.P. Shestopalov, Yu.V. Shestopalov. *Spectral theory and excitation of open structures.* Peter Peregrinus, London, (1996).
6. F.D. Gakhov. *Boundary value problems.* Nauka, Moscow, (1977) (in Russian).
7. I.Ts. Gohberg, M.G. Krein. *Introduction to the theory of linear nonselfadjoint operators in Hilbert space.* Nauka, Moscow, (1965) (in Russian).
8. I.S. Gradshtein, I.M. Ryzhik. *Tables of integrals, sums, series and products.* Fizmatgiz, Moscow, (1965) (in Russian).
9. G.A. Grinberg. Diffraction of electromagnetic waves by infinitely thin, perfectly conducting screens. *Zhurn. Tekhn. Fiziki.* **27**, 2326-2339 (1957).
10. G.A. Grinberg. A method for solving problems of electromagnetic wave diffraction by perfectly conducting plane screens based on the study of shadow currents. *Zhurn. Tekhn. Fiziki.* **28**, 542-568 (1958).
11. H.B. Dwight. *Tables of integrals and other mathematical data.* Macmillan, New York, (1961).
12. V.I. Dmitriev, E.V. Zakharov. *Integral equations in boundary value problems of electrodynamics.* Izd. Mosk. Gos. Univ., Moscow, (1987) (in Russian).
13. Yu.V. Egorov. *Linear differential equations of principal type.* Nauka, Moscow, (1984) (in Russian).
14. Yu.V. Egorov. *Lectures on partial differential equations. Additional chapters.* Izd. Mosk. Gos. Univ., Moscow, (1985) (in Russian).
15. Yu.V. Egorov, M.A. Shubin. Linear partial differential equations. Elements of modern theory. In: *Modern problems of mathematics. Fundamental trends.* **31**, 5-125, VINITI, Moscow, (1988) (in Russian).
16. Yu.A. Eremin, A.G. Sveshnikov. *Method of discrete sources in problems of electromagnetic diffraction.* Izd. Mosk. Gos. Univ., Moscow, (1992) (in Russian).

17. E.V. Zakharov, Yu.V. Pimenov. *Numerical analysis of radio wave diffraction.* Radio i Svyaz', Moscow, (1982) (in Russian).

18. A. Sommerfeld. *Vorlesungen uber theoretische Physik. Bd. 3. Elektrodynamik.* Akademische Verlagsgesellschaft, Leipzig, (1949).

19. A. Sommerfeld. *Vorlesungen uber theoretische Physik. Bd. 4.* Wiesbaden, (1950).

20. V.A. Il'in. On the convergence of expansions in eigenfunctions of the Laplacian. *Uspekhi Mat. Nauk.* **13**, 87-180 (1958).

21. V.A. Il'in. Localization and convergence of Fourier series in systems of fundamental functions of the Laplacian. *Uspekhi Mat. Nauk.* **23**, 61-120 (1968).

22. V.A. Il'in *Spectral theory of differential operators.* Nauka, Moscow, (1991) (in Russian).

23. A.S. Ilyinsky, V.B. Kravtsov, A.G. Sveshnikov. *Mathematical models of electrodynamics.* Vyssh. Shkola, Moscow, (1991) (in Russian).

24. A.S. Ilyinsky, Yu.G. Smirnov Integral equations for problems of diffraction on screens. *Radioteknika i Elektronika.* **39**, 23-31 (1994).

25. L.V. Kantorovich, G.P. Akilov. *Functional analysis.* Nauka, Moscow, (1984) (in Russian).

26. T. Kato. *Perturbation theory for linear operators.* Springer-Verlag, Berlin, (1972).

27. D. Colton, R. Kress. *Integral equation methods in scattering theory.* John Wiley, New York, (1983).

28. V.A. Kondrat'ev. Boundary value problems for elliptic equations in domains with conical and corner points. In: *Trudy MMO.* **16**, 209-292, Izd. Akad. Nauk SSSR, Moscow, (1967) (in Russian).

29. V.D. Kupradze. *Fundamental problems of mathematical theory of diffraction.* Gostekhizdat, Moscow, (1935) (in Russian).

30. V.D. Kupradze. *Boundary value problems of oscillation theory and integral equations.* Gostekhizdat, Moscow, (1951) (in Russian).

31. O.A. Ladyzhenskaya, N.N. Ural'tseva. *Linear and quasilinear equations of elliptic type.* Nauka, Moscow, (1973) (in Russian).

32. J.L. Lions, E. Magenes. *Problemes aux limites non homogenes et applications.* Dunod, Paris, (1968).

33. R. Mittra, S.W. Lee. *Analytical techiques in the theory of guided waves.* Macmillan, New York, (1971).

34. V.P. Mikhailov. *Partial differential equations.* Nauka, Moscow, (1983) (in Russian).

35. A.S. Mishchenko. *Vector bundles and their applications.* Nauka, Moscow, (1984) (in Russian).

36. S.P. Novikov, A.T. Fomenko. *Elements of differential geometry and topology.* Nauka, Moscow, (1987) (in Russian).

37. B.A. Plamenevskii. *Pseudodifferential operator algebra.* Nauka, Moscow, (1986) (in Russian).

38. M.M. Postnikov. *Smooth manifolds.* Nauka, Moscow, (1987) (in Russian).

39. N.P. Korneichuk. *Splines for approximation theory.* Nauka, Moscow, (1984) (in Russian).

40. S. Rempel, B.-W. Schulze. *Index theory of elliptic boundary value problems.* Akademie Verlag, Berlin, (1982).

41. V.M. Repin. Numerical methods for solving problems of electromagnetic coupling of volumes through a hole. *Zh. Vychisl. Mat. i Mat. Fiz.,* **11**, 152-163 (1971).

42. V.M. Repin. Diffraction of electromagnetic waves by a rectangular hole in a screen. In: *Computational methods and programming.* **24**, 34-48, Izd. Mosk. Gos. Univ., Moscow, (1975) (in Russian).

43. A.A. Samarskii, A.N. Tikhonov. On excitation of radio waveguides. Part 1. *Zhurn. Tekhn. Fiziki.* **17**, 1283-1296 (1947). Part 2. *Ibid.* 1431-1440 (1947).

44. A.A. Samarskii, A.N. Tikhonov. Representation of the field in waveguides in the form of the sum of TE and TM modes. *Zhurn. Tekhn. Fiziki.* **18**, 971-985 (1948).

45. E. Sanchez-Palencia. *Nonhomogeneous media and vibration theory.* Springer-Verlag, Berlin, (1980).

46. A.G. Sveshnikov. Radiation principle. *Dokl. AN SSSR.* **73**, 917-920 (1950).

47. A.G. Sveshnikov. Principle of limiting absorption for a waveguide. *Dokl. AN SSSR.* **80**, 345-347 (1951).

48. Yu.G. Smirnov. Fredholm property of the problem of diffraction by a planar, bounded perfectly conducting screen. *Dokl. AN SSSR.* **319**, 147-149 (1991).

49. Yu.G. Smirnov. *Solvabilily of vector electrodynamic problems on open surfaces.* Doct. Thesis, Moscow, (1995) (in Russian).

50. Yu.G. Smirnov. Solvability of integrodifferential equations in problems of diffraction by perfectly conducting planar screens. *Radioteknika i Elektronika.* **37**, 32-35 (1992).

51. Yu.G. Smirnov. Fredholm property of the system of pseudodifferential equations for the problem of diffraction by a bounded screen. *Differents. Uravn.* **28**, 136-143 (1992).

52. Yu.G. Smirnov. On the solvability of vector problems of diffraction in domains connected through an opening in a screen. *Comp. Maths Math. Phys.* **33**, 1263-1273 (1993).

53. Yu.G. Smirnov. *Application of Chebyshev polynominals to solving one-dimensional equations of potential type.* Izd. Penz. Gos. Tech. Univ., Penza, (1994) (in Russian).

54. Yu.G. Smirnov. Solvability of vector integrodifferential equqtions for the problem of the electromagnetic field diffraction by screens of arbitrary shape. *Comp. Maths Math. Phys.* **34**, 1265-1276 (1994).

55. V.G. Sologub. Short-wave asymptotic of solution to the problem of diffraction by a disk. *Zh. Vychisl. Mat. i Mat. Fiz.* **12**, 388-412 (1972).

56. V.G. Sologub. On the solution to an integral convolution-type equation with finite limits of integration. *Zh. Vychisl. Mat. i Mat. Fiz.* **11**, 637-654 (1970).

57. M.E. Taylor. *Pseudodifferential operators.* Princeton University Press, Princeton, (1981).

58. H. Triebel. *Interpolation theory. Function spaces. Differential operators.* Veb Deutshe Verlag, Berlin, (1978).

59. P.Ja. Ufimtzev. *Method of boundary waves in physical theory of diffraction.* Sov. Radio, Moscow, (1962) (in Russian).

60. Ja.N. Fel'd. *Fundamentals of the theory of slot antennas.* Sov. Radio, Moscow, (1948) (in Russian).

61. Ja.N. Fel'd. Diffraction of electromagnetic waves by open metal surfases. *Radioteknika i Elektronika.* **20**, 28-38 (1975).

62. H. Hönl, A.W. Maue, K. Westpfahl. *Theorie der Beugung.* Springer-Verlag, Berlin, (1961).

63. M.W. Hirsch. *Differential topology.* Springer-Verlag, Berlin, (1976).

64. B.V. Shabad. *Introducion to complex analysis.* Nauka, Moscow, (1969) (in Russian).

65. V.P. Shestopavlov. *Summation-type equations in the modern theory of diffraction.* Naukova Dumka, Kiev, (1983) (in Russian).

66. A.M. Shubin. *Pseudodifferential operators and spectral theory.* Nauka, Moscow, (1978) (in Russian).

67. G.I. Eskin. *Boundary value problems for elliptic pseudodifferential equations.* Nauka, Moscow, (1973) (in Russian).

68. N. Amitay, V. Galindo. On energy conservation and the method of moments in scattering problems. *IEEE Trans.* **AP-17**, 747-751 (1969).

69. M.G. Andreasen. Scattering from bodies of revolution. *IEEE Trans.* **AP-13**, 303-310 (1965).

70. T.S. Angell, G.C. Hsiao, J. Kral. Double layer potentials on boundaries with corners and edges. *Comment. Math. Unit. Carol.* **27**, 419 (1986).

71. M.C. Butler, K.R. Umanshankar. Electromagnetic excitation of a wire through an aperture-perforated conducting screen. *IEEE Trans.* **AP-24**, 456-462 (1976).

72. C.M. Butler, Y. Rahmat-Samii, R. Mittra. Electromagnetic penetration through apertures in conducting surfaces. *IEEE Trans.* **AP-26**, 82-93 (1978).

73. *Computational electromagnetics: frequency-domain method of moments.* Ed. by E.K.Miller, L. Medgyesi-Mitschand, E.H. Newman. IEEE Press, New York, (1992).

74. M. Costabel. Boundary integral operators on curved polygons. *Ann. Mat. Pura Appl.* **133**, 305-326 (1983).

75. M. Costabel. Boundary integral operators on Lipschitz domains: elementary results. *SIAM J. Math. Anal.* **19**, 613-626 (1988).

76. R.F. Harrington. *Field computation by moment methods.* Macmillan, New York, (1968).

77. R.F. Harrington, J.R. Mautz. Computational methods for transmission on waves. In: *Electromagnetic Scattering,* 429-470. Academic Press, New York, (1978).

78. J.J. Kohn, L. Nirenberg. An algebra of pseudodifferential operators. *Commun. Pure and Appl. Math.* **18**, 269-305 (1965).

79. W. Magnus, F. Oberhettingen. Uber einige Randwertprobleme der Schwihgungsgleichung $\Delta u + k^2 u = 0$ im Falle ebener Begrenzungen. *J. Reine Angew. Math.* **186**, 184-192 (1949).

80. A.W. Maue. Toward formulation of a general diffraction problem via an integral equation. *Zeitschrift fur Physik.* **126**, 601-618 (1949).

81. E.K. Miller, A.J. Poggio. Moment-method techniques in electromagnetics from applications viewpoint. In: *Electromagnetic Scattering,* 315-358. Academic Press, New York, (1978).

82. *Numerical and asymptotic techniques in electromagnetics.* Ed. by R. Mittra. Springer-Verlag, Berlin, (1975).

83. J. Moore, R. Pizer. *Moment methods in electromagnetics: techniques and applications.* John Wiley, New York, (1984).

84. J.A. Morrison, J.A. Lewis. Charge singularity at the corners of a flat plate. *SIAM J. Appl. Math.* **31**, 233-250 (1976).

85. K. Morgenröther, P. Werner. On the instability of resonances in parallel-plane waveguides. *Math. Meth. in the Appl. Sci.* **11**, 279-315 (1989).

86. Cl. Müller. *Foundations of the mathematical theory of electromagnetic waves.* Springer-Verlag, Berlin, (1969).

87. L. Päivärinta, S. Rempel. A deconvolution problem with kernel $1/|x|$ on the plane. *Appl. Anal.* **26**, 105-128 (1987).

88. L. Päivärinta, S. Rempel. Corner singularities of solutions to $\Delta^{\pm 1/2} u = f$ in two dimensions. *Asymptotic Analysis.* **5**, 429-460 (1992).

89. F.G. Ramm. *Scattering by obstacles.* D. Reidel Publ. Comp., Dordrecht, (1986).

90. S.M. Rao, D.R. Wilton, A.W. Glisson. Electromagnetic scattering by surfaces of arbitrary shape. *IEEE Trans.* **AP-30**, 409-418 (1982).

91. H.K. Schuman, D.E. Warren. Aperture coupling in bodies of revolution. *IEEE Trans.* **AP-26**, 778-783 (1978).

92. Yu.G. Smirnov. Pseudodifferential equations for electrodynamic screen problem in R^3. In: *Mathematical methods in electromagnetic theory. Proc. of the 4th International Seminar. September 15-24 , Alushta,* 171-182. Test-Radio, Kharkov, (1991).

93. Yu.G. Smirnov. The behavior of electromagnetic field near a corner of a flat bounded screen. In: *Day of Diffraction-92 International Seminar. Abstracts. St Petersburg,* 31 (1992).

94. E.P. Stephan. Boundary integral equations for screen problems in R^3 . *Integral Equations and Operator Theory.* **10**, 236-257 (1987).

95. E.P. Stephan, W.L. Wendland. An augmented Galerkin procedure for the boundary integral method applied to two-dimensional screen and crack problems. *Applicable Analysis.* **18**, 105-128 (1984).

96. I. Toyoda, M. Matsuhara, N. Kumagai. Extended integral equation formulation for scattering problems from a cylindrical scatterer. *IEEE Trans.* **AP-36**, 1580-1586 (1988).

97. R.F. Wallenberg, R.F. Harrington. Radiation from an aperture in conducting cylinders of arbitrary cross section. *IEEE Trans.* **AP-17**, 56-62 (1969).

98. J.H.H. Wang. *Generalized moment methods in electromagnetics.* John Wiley, New York, (1991).

99. D.R. Wilton, C.M. Butler. Effective methods for solving integral and integrodifferential equations. *Electromagnetics.* **1**, 289-308 (1981).